'Martyr of Science':
Sir David Brewster 1781–1868

Frontispiece Sir David Brewster. Engraved by D. J. Pound from a photograph by John Watkins. *(Scottish National Portrait Gallery).*

'Martyr of Science':
Sir David Brewster 1781–1868

*Proceedings of a Bicentenary Symposium
held at the Royal Scottish Museum on 21 November 1981,
together with a catalogue of scientific apparatus
associated with Sir David Brewster
and a bibliography of his
published writings.*

Edited by A. D. Morrison-Low and J. R. R. Christie

Royal Scottish Museum Studies

The Royal Scottish Museum
Edinburgh 1984
© Crown copyright 1984
ISBN 0 900733 29 2

British Library Cataloguing in Publication Data
Martyr of science.—(Royal Scottish Museum studies)

1. Brewster, *Sir* David 2. Scientists—Scotland
—Congresses
I. Morrison-Low, A. D.   II. Christie, J. R. R.
III. Series
509'.2'4   Q143.B/

Foreword

During his lifetime, Sir David Brewster was recognised as one of the country's outstanding scientists, whereas today his is scarcely a household name, and he is perhaps remembered only as the inventor of that delightful childhood toy, the kaleidoscope. However, the bicentenary of his birth provided a suitable occasion to re-examine some of the positive contributions made by Brewster, not just to science, but to the social and cultural history of the nation.

It was appropriate that this symposium, which celebrated a man whose research work ranged from experimental science through biology to the geological sciences, was held in the national museum in his native country, an institution which reflects the diversity of Sir David's own scientific interests.

I should like to offer our thanks for great help and kind support to a number of bodies who have been associated with this anniversary meeting: the Universities of St. Andrews and Edinburgh, at which Brewster held consecutive posts as Principal; the Royal Society of Edinburgh, of which he was for many years Secretary, and at his death President; the Royal Scottish Society of Arts, which he established in 1821; the British Association for the Advancement of Science, which he helped found some ten years later; and the British Society for the History of Science, a much younger body than Sir David's two hundred years, which maintains great interest in all things historical and scientific.

Sir David's anniversary gives us a timely opportunity to review his diversity. So diverse was he, indeed, that a mere day's symposium could not hope to cover all the spheres in which he was influential. We hope that an occasion such as this will not be seen merely as an end in itself, but as a starting point for further studies, and a vignette of Brewster's life in a more critical historical context.

NORMAN TEBBLE DSc
DIRECTOR
ROYAL SCOTTISH MUSEUM

Preface

David Brewster has always been a difficult figure to characterise for historians of science. His life and work have so far not drawn the concentrated and systematic scholarly attention devoted to the achievements of such glittering contemporaries as Maxwell or Darwin. Yet though Brewster might never have produced a single piece of work which compares with theirs, in recent years he has generated increasing interest among scholars whose attention has shifted from narrowly-based scrutiny of great works in the history of science, towards a wider-ranging consideration of science's social, political and cultural context in nineteenth-century Britain. Once such a shift of scholarly focus had occurred, Brewster was bound to emerge as a much more complex and interesting figure than had hitherto been thought. The widely varied nature of his career, the breadth of his intellect and his natural combativeness all combined to place him at the intersection of several increasingly important topics currently under investigation. These include such developments as the explosive growth of scientific publishing, the origination of novel forms of institutionalisation for science, and deeply-rooted conflicts over the relations of science and religion, to mention only the most obvious. Brewster's activities and his evangelism in science and religion kept him in the forefront of these developments, while his ready pen and willingness to engage in intellectual confrontation and polemic meant that on these issues, and on others to do with his own areas of special expertise and interest (optics, optical instrumentation and the history of science), he was a difficult man for contemporaries, and now for historians, to ignore.

Given these reasons for a renewed interest in Brewster, the obvious and main conclusion is that a re-examination of him is now due. The most effective form for this re-examination to take will be, in the longer term, a full-scale biographical study which would pursue in detail all the variegated aspects of Brewster's extraordinarily busy and productive life. Such a study could prove extremely revealing, not only of the individual himself, but of many of the complex ramifications of British science between 1800 and 1860.

The present collection of essays does not pretend to the scope which such a biography would possess. It gathers together a number of studies whose common focus is Brewster, but Brewster as seen, to employ a Brewsterian metaphor, through a series of different lenses: his technical interests in instrumentation and optics; the institutional and social worlds of the Scottish Universities, Edinburgh science and the British Association; his involvement in the precarious commercial world of scientific journalism and publishing; his deeply committed Evangelicalism; and his enthusiastic, puzzled encounters with the lives of past scientists. Each essay throws new light on the ways in which Brewster worked, thought and wrote, and it is our hope that, collectively, they may stimulate and to an extent direct further interest in and research on David Brewster and the burgeoning world of science he so memorably inhabited.

We would like to thank the following for their contributions to the planning and execution of the Brewster Bicentenary Symposium and Exhibition: Symposium session chairmen, Professor David Daiches, Professor Eric Forbes, Dr. Charles Waterston, Dr. N. T. Phillipson; colleagues in the Royal Scottish Museum; and the staff of the Royal Society of Edinburgh, for their part in the organisation and administration of the Symposium.

For their help and advice in lending material for the Exhibition, we wish to thank: Alan Baxter W.S., Mrs. M. Brewster-Macpherson, Alan Fletcher, W. H. Veitch; staff of the British Museum, the Science Museum, London, Hertford Museum, and the University Library and Department of Physics at the University of St. Andrews; Patrick Cadell of the National Library of Scotland, for all his help with the second exhibition of Brewster manuscripts displayed after the Symposium; and Miss Sara Stevenson of the Scottish National Portrait Gallery.

Finally, we would like to thank all the speakers for their efforts, both on the day of the Symposium itself and afterwards; and an especially grateful vote of thanks to Dr. A. D. C. Simpson for his support and wisdom before, during and after the Symposium, and his help with the publication of this volume.

A. D. MORRISON-LOW

ROYAL SCOTTISH MUSEUM

J. R. R. CHRISTIE

UNIVERSITY OF LEEDS

Contributors

Dr. Robert Anderson
Lecturer
Department of History
University of Edinburgh

Paul Baxter
Senior Scientific Officer
The British Library
Research and Development Department
London

Dr. W. H. Brock
Director
Victorian Studies Centre
University of Leicester

Patrick Cadell
Assistant Keeper
Department of Manuscripts
National Library of Scotland

Dr. G. N. Cantor
Lecturer
Division of the History and Philosophy of Science
University of Leeds

John R. R. Christie
Lecturer
Division of the History and Philosophy of Science
University of Leeds

Professor William Cochran
Chair of Natural Philosophy
Department of Physics
University of Edinburgh

Professor David Daiches
(Chairman)
Director
Institute for Advanced Studies in the Humanities
University of Edinburgh

Dr. George Duncan
Reader in Biophysics
School of Biological Sciences
University of East Anglia

Professor Eric Forbes
(Chairman)
Chair in History of Science
History of Medicine and Science Unit
University of Edinburgh

J. B. Morrell
Lecturer in History of Science
School of Social Sciences
University of Bradford

Miss A. D. Morrison-Low
Assistant Keeper
Department of Technology
Royal Scottish Museum

Dr. Nicholas Phillipson
(Chairman and speaker)
Senior Lecturer
Department of History
University of Edinburgh

Dr. Steven Shapin
Lecturer
Science Studies Unit
University of Edinburgh

Dr. Charles Waterston
(Chairman)
Keeper
Department of Geology
Royal Scottish Museum

Contents

page

11 Sir David Brewster: An Outline Biography
Professor William Cochran

17 Brewster and the Edinburgh Career in Science
Dr. Steven Shapin

25 Brewster and the early British Association for the Advancement of Science
J. B. Morrell

31 Brewster and the Reform of the Scottish Universities
Dr. Robert Anderson

37 Brewster as a Scientific Journalist
Dr. W. H. Brock

45 Brewster, Evangelism and the Disruption of the Church of Scotland
Paul Baxter

53 Sir David Brewster as an Historian of Science
John R. R. Christie

59 Brewster and Scientific Instruments
A. D. Morrison-Low

67 Brewster on the Nature of Light
Dr. G. N. Cantor

79 Sir David Brewster: some concluding remarks
Dr. Nicholas Phillipson

Appendix I

82 a. Scientific Apparatus associated with Sir David Brewster: An Illustrated Catalogue
A. D. Morrison-Low

101 b. Brewster's Contribution to the Study of the Lens of the Eye: An Experimental Foundation for Modern Biophysics.
Dr. G. Duncan

Appendix II

105 a. Papers of Sir David Brewster in the National Library of Scotland
Patrick Cadell

107 b. Published Writings of Sir David Brewster: a Bibliography
A. D. Morrison-Low

137 c. Published Writings about Sir David Brewster
A. D. Morrison-Low

Sir David Brewster: An Outline Biography

WILLIAM COCHRAN

I first became interested in Sir David Brewster while reading about James Clerk Maxwell. I came across a reference to Maxwell as a very young man, getting up at a meeting of the British Association to challenge the formidable Sir David Brewster on a point connected with the theory of colour[1]. On further investigation Brewster impressed me sufficiently that the title of the talk I was preparing became 'Clerk Maxwell and other Edinburgh physicists'. Our main source of information about Brewster is *The Home Life of Sir David Brewster*, by his daughter, Mrs. Gordon[2]. Fortunately the word 'home' was not narrowly interpreted by the lady novelist, and all of his activities, including to some extent his science, are covered. It is not a hagiography; she did not hide the fact that her distinguished father could be irritable, impatient, litigious and verbally aggressive, and he emerges as a man with a strong personality, strong convictions and great personal charm when he chose to exercise it. While I have a limited amount of space, Mrs. Gordon required nearly 500 pages to describe the life of a man who was at various times a student of divinity, tutor, writer on popular science, experimental physicist, inventor, manager of an estate in the Highlands, university principal and reformer, crusader for science, historian of science and advocate of social reform. Above all he emerges from his daughter's account as a man who was the very model of a Scottish Presbyterian, who would admit no possibility of conflict between science and religion and who firmly believed that the progress of science could only benefit mankind.

In 1965 a reprint of Brewster's *Memoirs of the Life, Writings, and Discoveries of Sir Isaac Newton* was published[3], and it includes an introduction on Brewster by R. S. Westfall, himself a recent biographer of Newton. From Mrs. Gordon's broad canvas Westfall draws a lively sketch and, more importantly, he makes a modern assessment of Brewster's work. For Brewster, the historian of Newton, he is full of praise, but on Brewster the scientist his judgement is different — '... Brewster confined himself to filling in the minutiae of the existing system. He measured the refractive and dispersive powers of nearly 200 substances. He examined the optical properties of countless crystals ... Fresnel ... created a revolution in optics, Brewster conformed his work to the existing system and created the kaleidoscope'[4].

Many of Brewster's contemporaries, to say nothing of Brewster himself, would have been surprised at such a summing up. 'He is, in fact, at the head of our men of science for the originality of his researches and the importance of his discoveries.' That was written by Lord Brougham[5], a former Lord Chancellor and himself an amateur scientist. Admittedly he had been a close friend of Brewster for about 60 years when he penned the sentence. William Whewell, Master of Trinity College, Cambridge, could only be suspected of bias in the other direction, since he and Brewster clashed more than once. In his *History of the Inductive Sciences* Whewell described Brewster as 'the Kepler of optics', and in a second edition he added that 'the immense number and variety of the beautiful optical discoveries which we owe to Sir David Brewster makes the comparison in his case a very imperfect representation of his triumphs over nature'[6]. Sir George Airy, then Astronomer Royal, described Brewster simply as 'the father of modern experimental optics'[7]. (Again it must be admitted that in private they appear to have been sometimes less kind.)

In his own lifetime he was accorded a degree of recognition as great as for any other natural philosopher, before or since. As well as a knighthood, he received many honorary degrees, almost every one of the medals in the gift of the Royal Society and of the Royal Society of Edinburgh, and he was an

Fig. 1. Sir David Brewster, his daughter Constance and an unknown woman, at Allerly near Melrose *c.*1865. (*Mrs. M. Brewster-Macpherson*).

honorary member of so many foreign academies that he always concluded the list with at least one '*etc*'. A new Australian plant was named *Cassia Brewsteri*, a mineral was named Brewsterite and a promontory in Greenland became Cape Brewster.

What then *were* Brewster's principal contributions to physics? That is after all a part of his biography. Surprisingly, Westfall did not even mention Brewster's Law, by which his name is memorialised. Almost all of Brewster's more important discoveries were concerned with polarised light, and his most innovative period was between about 1811 and 1818, although he continued to experiment with light and its interaction with materials of animal, vegetable and mineral origin for another fifty busy years.

When a ray of light falls on a plane face of a crystal such as calcite, an 'ordinary' ray travels into the crystal in accord with Snell's law of refraction, but an 'extraordinary' ray is also produced and takes a direction different by at most a few degrees. The crystal is said to be doubly refracting. There is one direction in the crystal, determined by its symmetry and called the optic axis, for which the two rays coincide. In 1678 Huygens, using a wave theory of light, or more correctly a pulse theory, was successful in finding a geometrical construction which gives the paths of both rays, but he was not able to decide what property it is that distinguishes the refracted rays from one another or from the incident ray. We now know that the distinction is that while the electric field of the incident light has no preferred direction of vibration within a plane, the ordinary ray has its vibrations confined to one direction, while those of the extraordinary ray are in a perpendicular direction. There was little progress on the subject for a century following Huygens' work until in 1808 Malus made a chance observation which led him to the discovery that light reflected at a particular angle from glass has the same property as a ray formed by refraction in calcite. Malus coined the term 'plane polarised' to describe the reflected light.

Brewster had apparently an incomplete account of this work—the Napoleonic wars were in progress. He repeated Malus's observations and showed empirically that the angle of incidence ϕ for which the polarisation is complete is given by $\tan \phi = n$, where n is the refractive index of the reflecting medium. This is Brewster's Law, which would have ensured him a place in the history of optics even if he had done no more. He also investigated the polarisation of the refracted light, where he was unknowingly repeating Malus's observations. The analysis of the polarised light was by subsequent reflection, with a different configuration.

Using extremely simple apparatus he made another important discovery, and the fact that Brewster made it has been completely forgotten. When a thin slice of a doubly-refracting crystal is placed between the polariser and the analyser, and when an extended source is used and the scale of the apparatus reduced so that a convergent beam of light reaches the observer's eye, beautiful optical figures are seen. Nowadays a polarising microscope would be used. The circles on a maltese cross surround the optic axis, in calcite for example. In the mineral topaz Brewster saw two systems of circles; in potassium nitrate, for example, they were close enough to be seen at the same time. This led him to the conclusion that certain crystals have two optic axes, and exhibit a more complicated type of double refraction than had been studied by Huygens in calcite. After months of observations on 'countless crystals'[8], Brewster arrived at an important generalisation in which he related optical properties to crystal symmetry—I shall not give the details but in my view this was his most important contribution. Text books of optics and of crystallography alike fail to mention that it was Brewster who made almost all the observations, and who arrived at a valid generalisation.

His other work included the discovery that an optically isotropic substance such as glass becomes doubly refracting when mechanically strained, or when strained by unequal heating. His work on the structure of the lens of the eye, which he began about 1816 and to which he several times returned, confirmed the existence of an ordered fibrous structure in the lens. He was able to measure the periodicity of the structure using a diffraction technique with a candle as source, and according to Dr. G. Duncan[9] his data has scarcely been improved on in recent work using a laser as source. His theory of colour and of the nature of the spectrum of white light was overturned by the work of Helmholtz and of Maxwell, but he never abandoned it.

It was not until 1817 that Young realised how polarisation could be accommodated in a wave theory of light. Fresnel arrived independently at the same conclusions, and it was largely his genius that took the wave theory to the point where it accounted for all the facts concerning reflection, refraction, diffraction, interference, polarisation and double refraction, and could be used to predict new phenomena. Brewster's attitude to the 'undulatory theory' was however at best agnostic; he avoided using it and he did not expound it in his *Treatise on Optics* where he merely noted that '... the undulatory theory has been received by many of our most distinguished philosophers and adopted even by those who do not admit it as a physical truth'[10].

We can I think see one reason why his reputation has declined. He seldom interpreted his observations within an acceptable theoretical framework; this was done by others and the connection with Brewster has sometimes been lost. I do not have time to pursue this topic further, nor can I do more than indicate the range of his curiosity and the extent of his industry as a scientist—Mrs. Gordon lists 315 papers published between 1806 and 1868, and that does not include his books, or his articles intended for a wider public.

I turn now to the story of his life, where my task, as I see it, is to put some sign-posts to guide readers from one paper to the next. Brewster's father was the rector of Jedburgh School, and parental influence pointed him towards the Church, in which his three brothers found careers. Another formative influence during his schooldays was his friendship with James Veitch, in whose workshop he spent much of his spare time. Veitch was a ploughwright who also made scientific instruments. Sir Walter Scott described Veitch as '... a self-taught philosopher, astronomer and mathematician ... and certainly one of the most remarkable persons I ever knew'[11]. With Veitch's help, Brewster constructed his own telescope at the age of ten. There is a drawing of the Schoolmaster's house in Jedburgh, sketched about 1829 on the back of an envelope by James Forbes[12], of whom I shall have more to say later. A note indicates the window from which Brewster as a schoolboy tested his telescope. As a student of divinity at Edinburgh University, Brewster kept in touch with Veitch and they corresponded about both religious and scientific matters. Brewster's professors included Dugald Stewart (moral philosophy), John Playfair (mathematics) and John Robison (natural philosophy). At university Brewster began a life-long friendship with Henry Brougham, later Lord Brougham. They shared an interest in optics and separately made experiments in their students days on the diffraction of light, without however achieving anything very novel[13].

Brewster was some time in finding a career. Although a devoted adherent of the Evangelical wing of the Church of Scotland, his dislike of public speaking and inability to give a sermon from memory prevented him from becoming a parish minister. He was for some years a resident tutor with well-to-do families in the neighbourhood of Edinburgh, and gradually he found that he could make a modest living by means of his pen. He was prepared to write, and write well, on a great variety of topics—archaeology, colour-blindness, meteorology, education, and so on. As Mrs. Gordon expressed it, 'rapidly his occupied mind poured itself upon paper'[14]. He became editor of the *Edinburgh Magazine* at the age of 20, and he embarked on the project of bringing out an encyclopaedia, for which he wrote many of the articles himself. For over 30 years he supported himself and eventually a growing family by 'the precarious profession of literature'[15], and from about 1806 he found time for research, at first on optical instruments and later, also on polarised light. Seldom has the Presbyterian work-ethic been better displayed—although Mrs. Gordon hints at a certain lack of organisation in his efforts, and he was never the man to fall into routine ways.

It was possibly financial insecurity which prompted his concern with the status of scientists and inventors. 'England alone taxes inventors as if they were the enemy of the state', he wrote[16]. In the 1820s he shared the view that science was in decline nationally, and one result of the campaign which was mounted was the foundation of the British Association[17]. Brewster was never financially astute, and his inventions, including the kaleidoscope and the polyzonal lens, brought him no profit, according to Mrs. Gordon[18]. In the early 1830s, threatened by a protracted law-suit involving his encyclopaedia, he had to consider desperate measures. Lord Brougham offered to find him a living in the Church of England, but the plan fell through. In 1832 he applied for the chair of natural philosophy in Edinburgh University. Only financial necessity can explain this step, for he had earlier written to his young friend and protégé James Forbes advising him to stick to the law as a profession: 'There is no profession so incompatible with original inquiry as a Scotch Professorship, where one's income depends on the number of pupils. Is there one Professor in Edinburgh pursuing science with zeal? Are they not all occupied as showmen whose principal object is to attract pupils and make money?'[19] He mentioned incidentally, that he had given the same advice to Thomas Carlyle, successfully deflecting him from science to literature. Ironically it was Forbes who was Brewster's only serious rival for the chair, and after an undignified campaign on Forbes's part, it was he who was appointed. It was possibly a correct decision for wrong reasons, and as a mutual friend wrote to Forbes, 'I cannot but feel surprised at his *seeking* an appointment for which age, temperament and habits so manifestly unfit him. Everyone who knows what it is to lecture to a class of persons ... full of rampant animal spirits ... will be able to conceive the intolerable annoyance of such conduct to a sensitive person like our excellent friend Sir David'[20]. I am specially interested in his non-appointment since the chair is, metaphorically speaking, the one in which I now sit.

For a few years Brewster made ends meet by managing an estate, Belleville, (now Balavil), near Kingussie, which his sister-in-law had just inherited indirectly from her father, James Macpherson, better known as 'Ossian' Macpherson, creator/discoverer of certain Gaelic epic poetry[21]. In 1838, at the age of fifty-six, Brewster received for the first time a regular appointment, as a Principal at the University of St. Andrews. (I cannot help remarking that the advantages of *retirement* at fifty-five from a university post are now being advocated). The post was in those days something of a sinecure, but Brewster was not the man to treat it as such. I shall leave it to Dr. Robert Anderson to tell you this chapter of the story[22]; suffice it to say that Brewster's intemperate pressure for reform soon had the place in an uproar, and in 1843 his professors attempted to have him deprived of office, on the specious grounds that he was no longer a member of the Established Church. When he did resign, much later, it was of his own accord, and he left the University a better place than he had found it.

Life begins at forty, many of us have been told. One November day Brewster was travelling by coach on his way to a winter holiday in Cannes when he got into conversation with a young lady, one of a party of three on their way to holiday in Nice. Soon after his arrival Brewster decided that he preferred Nice to Cannes, and after a whirlwind courtship he and Miss Jane Purnell were married and left for a lengthy tour of Italy. My remarks may have given you the impression that Brewster was then about forty, but in fact he was within a few weeks of his seventy-fifth birthday when he met his second wife[23]. It was a happy marriage, according to Mrs. Gordon, who may have been older than her step-mother. Brewster continued his new life by resigning from St. Andrews, *not* in order to go into retirement but because Edinburgh had decided to make amends, and in 1859 he became Principal of that University. One of his first official duties was to preside over the installation of his friend Lord Brougham as Chancellor. Brewster's eight years here were comparatively peaceful ones; the pace of life was slower than it is now but the trains were better and he was able to commute from Melrose. After his death he was commemorated by a statue in the Old Quad, the only Principal to be honoured in this way. There is surely something symbolic in the fact that the statue now stands before the Chemistry Department, while a short distance away the Physics Department is housed in the James Clerk Maxwell building. It is not for me to sum up. There are aspects of his life which I have scarcely mentioned and which I must leave entirely to other contributors to this Symposium.

I shall allow Westfall space to make amends and give him the last word. 'He deserves to be restored in memory to that prominent role which by sheer force of personality he played in fact. He deserves to be remembered as one of humanity's prouder exhibits. What an inexhaustible reservoir of vitality! What a fund of human sympathy!'[24]

Notes and References
I am grateful to Dr. A. D. C. Simpson, Miss A. D. Morrison-Low, Mr. A. S. McKirdy, Mr. R. N. Smart, Dr. H. Montgomery and Dr. G. Duncan for bringing to my attention material which I would otherwise have missed.

1. L. Campbell and W. Garnett, *The Life of James Clerk Maxwell* (London, 1882), 144.
2. [Margaret Maria Brewster] Gordon, *The Home Life of Sir David Brewster* (1st edition, Edinburgh, 1869).
3. D. Brewster, *Memoirs of the Life, Writings, and Discoveries of Sir Isaac Newton*. (Reprinted from the Edinburgh edition of 1855: New York, 1965).
4. R. S. Westfall, 'Introduction', *op. cit.* (3), ix–xlv.
5. Quoted by Gordon, *op. cit.* (2), 381.
6. W. Whewell, *History of the Inductive Sciences* 3 vols. (3rd edition, London, 1857) II, 370–371.
7. Quoted by Whewell, *op. cit.* (6), 373.
8. Westfall, *op. cit.* (4), xviii.
9. G. Duncan, 'Brewster's Contribution to the Study of the Lens of the Eye: An Experimental Foundation for Modern Biophysics', in this volume.
10. D. Brewster, *A Treatise on Optics* (2nd edition, London, 1853), 164. It is worth noting that in the 1st edition of 1831, p. 135, the sentence quoted ended at the word 'philosophers'.
11. Quoted by Gordon, *op. cit.* (2), 25.
12. The correspondence of J. D. Forbes is held in the Library of the University of St. Andrews. The drawing is on blank pages of a letter from Brewster to Forbes dated August 1st, 1829.
13. This remark should be modified in the light of Dr. G. N. Cantor's paper, 'Brewster on the Nature of Light', in this volume.
14. Gordon, *op. cit.* (2), 247.
15. Gordon, *op. cit.* (2), 95.
16. Quoted by Gordon, *op. cit.* (2), 148.
17. J. B. Morrell, 'Brewster and the Early British Association', in this volume.
18. A. D. Morrison-Low has found circumstantial evidence which indicates that Brewster did benefit financially from his invention of the kaleidoscope, but perhaps not to the extent for which he had hoped. Morrison-Low, 'Brewster and Scientific Instruments', in this volume.
19. Brewster to Forbes, *loc. cit.* (12), 11 February, 1830.
20. The mutual friend was William Henry (1774–1836). Henry to Forbes, *loc. cit.*, (12), 8 December, 1832.
21. B. Saunders, *The Life and Letters of James Macpherson* (2nd edition, London, 1895).
22. Robert Anderson, 'Brewster and the Reform of the Scottish Universities', in this volume.
23. Brewster's journal of his expedition is in the Museum, Queen Mary's House, Jedburgh. It was presented by Constance Hope (Brewster's daughter by his second marriage), about 150 years after his birth!
24. Westfall, *op. cit.* (4), xxx–xxxi.

Brewster and the Edinburgh Career in Science

STEVEN SHAPIN

In 1830 David Brewster offered his young protégé James David Forbes some free advice:

'I cannot tolerate [Brewster said] the idea of a Professorship being an object of your ambition, if you mean a Scotch one. There is no profession so incompatible with original inquiry as a Scotch Professorship, where one's income depends on the number of pupils'.

Legitimate scientific reputation, Brewster went on, is to be secured from research, not from teaching 'boys and half-men the elements of Euclid'[1]. Later in the year, perhaps with young Forbes in mind, Brewster elaborated his reservations about the professorial life:

'A small number of chairs in our universities are certainly the only rewards which are open to scientific ambition... Few, however, though they be, they will operate as an excitement to the young philosopher in his sacrifice of all other professional expectations; but the benefit thus conferred upon science is, in our opinion, far outweighed by the baneful influence which such situations produce on the philosopher who obtains them... No sooner is a professor installed behind the counter of his lecture-room, then it becomes his single object to enrich himself with the fees of his ready-money customers. His handbills announce the qualities of his wares;—the cups and balls and the fire-works of science are summoned into requisition, and by the legerdemain and alchemy of his art he transmutes his baser metals into gold'[2].

Perhaps Brewster had vividly in mind the example of Edinburgh professor of chemistry Thomas Charles Hope who committed the unpardonable sin of pandering to the ladies more effectively than Henry Cockburn. Cockburn retaliated by leaving to posterity one of his most devastating pen-portraits:

'Each of [the ladies in Hope's popular classes] brings a beau, and the ladies declare there was nothing so delightful as these chemical flirtations. The Doctor is in absolute extacy with his audience of veils and feathers, and can't leave the Affinities. Horrible – – –. I wish some of his experiments would blow him up. Each female student would get a piece of him'[3].

Or perhaps Brewster was thinking of John Leslie's even more bizarre strategy for grabbing his students' attention. 'Among other foibles', the *Edinburgh Weekly Chronicle* informed its readers, 'he used to stain his hair, which was sometimes blue, purple, yellow, or all the colours of the rainbow at once'—Scotland's first, and perhaps last, punk physicist[4].

Brewster's characterisation of the Edinburgh scientific professor as entrepreneurial charlatan, while extreme, was not without substance. As Jack Morrell has shown, it was only by appalling overwork that a commitment to scientific research could be meshed with the exigencies of practical pedagogy and the necessity of securing a satisfactory income through class-fees[5]. T. C. Hope opted for popularity and did little original research. John Leslie tried desperately to follow Hope's example, but, despite liberal use of the dye-bottle, could never match the chemist's audience of up to 500 at three or four guineas a head. Other Edinburgh scientific professors, like the natural historian Robert Jameson, did manage to combine active research with a decent popularity and income, but, in general, Brewster's point is well-taken. The Edinburgh scientific professor in the first decades of the nineteenth century was employed to teach science, not to produce science.

Thus, it should not come as a complete surprise that Brewster counselled young Forbes that the career most suited to the active prosecution of scientific research was that of the law. A well-managed legal life leading to, perhaps, a county sheriffship, might leave

Fig. 2. (a) and (b) The Keith Medal, awarded to David Brewster by the Royal Society of Edinburgh. *(A. G. L. Baxter, W.S.)*.

'two-thirds of your time to science'[6]. Forbes resisted Brewster's advice and countered with the suggestion of an author's life. Again, Brewster could not agree: 'I do not object to your making money by your writings, but I am sure that it would be injurious to your happiness to rely on such a source for a permanent portion of your income. The moment you do that you become a professional author, following the worst of all professions'[7].

Forbes could scarcely be blamed for not taking Brewster's advice, for Brewster did not take it either. Within two years of giving Forbes the benefit of his wisdom Brewster and Forbes were locked in a contest for the Edinburgh chair of natural philosophy, left vacant by John Leslie's death[8]. Forbes was victorious, and, until 1838, when at the age of 57 he became Principal of St. Andrews, Brewster was obliged to combine 'the worst of all professions' with that of harassed editor, under-rewarded inventor, and general factotum to Edinburgh scientific institutions.

The contest between Forbes and Brewster in 1832–33 is especially revealing of the obstacles Brewster faced in securing a career in science. Despite his misgivings, this was not the first time that Brewster thought seriously about a Scottish scientific chair. In 1805 he apparently considered applying for the Edinburgh mathematics chair that John Leslie succeeded in winning after a bitter contest. Then in 1807 he failed to win the mathematics chair at St. Andrews[9]. In 1816 he was made a firm offer by the Lord Provost to teach John Playfair's natural philosophy class at Edinburgh, a role which would have put Brewster in line of succession. Brewster declined, saying that he had no time to prepare for the lectures adequately and that he could not do the job 'without putting to hazard both my health and my reputation'. He was at that period more than usually frantic about the state of the *Edinburgh Encyclopaedia* and risked 'the utter ruin of my own private affairs' if he did not give all his attention to that business[10]. Then in 1828 he expressed great interest in the sinecure Edinburgh chair of practical astronomy, even going so far as to solicit testimonials from English astronomers. But he worried that the salary attached even to this non-teaching position would be insufficient for his family's needs, and, when it proved to be only £120 a year, he withdrew[11]. The 1816 episode in particular came back to haunt Brewster, for in the contest with Forbes a whispering campaign was set on foot which put it about, so Brewster said, that 'I was unable to lecture for Mr. Playfair in 1816; ... that ill health still disqualifies me; ... and that, if I did receive the appointment, I should require the aid of an assistant'[12]. These rumours owed much of their plausibility to Brewster himself, for in 1830 he had publicly suggested a new structure for scientific professorships. Let there be established, he wrote, chairs 'for the maintenance of men of genius, whose duty should be limited to the advancement of science by their original researches' and the instruction of unusually talented young men, while the drudgery of teaching the scientific proletariat might be left 'to a popular deputy appointed either by the professor or by the patron'[13].

Of course, the issue uppermost in the minds of the Edinburgh Town Council was Sir David's capacity to deliver public lectures. This was a legitimate concern, given the nature of professorial duties at the time. Brewster's 'anxious nervous temperament' failed him throughout his career. He only just managed to get through his first sermons as a minister in 1804, and, as his daughter related, 'never forgot the intense, though suppressed, suffering of the effort'. The strain of public performances even afflicted him when asked to say grace at dinner parties. On one memorable occasion, 'He began, but as he went on the words choked in his mouth, and he sat down in a faint'[14]. It was this extreme nervousness, rather than any lack of religious fervour, which decided Brewster on giving up the ministry as a career. But his problem continued to plague him in the world of science. Of his many papers for the Royal Society of Edinburgh, few, if any, were actually read by Brewster himself. In providing testimonials for the Edinburgh natural philosophy chair Brewster's friends tried to gloss away his affliction. The Evangelical minister Thomas Guthrie Wright blandly noted that Brewster could *write* lectures, 'and what reason is there for supposing that he cannot *read* them, or exhibit the usual experiments?' Brewster's complaint was put down to his great 'modesty'[15]. There were, however, occasions on which Brewster shone, especially when he was speaking about his own experimental work, before an audience he respected and which he knew respected him. Thus Vernon Harcourt's testimonial reported that Brewster's *extempore* performances at the 1831 York meeting of the British Association were 'delivered with perfect facility and fluency'; in similar terms J. F. W. Johnston applauded Brewster's 'fluency and felicity of expression'; and John Phillips commended his 'clear, complete, and impressive delivery'[16]. Unfortunately for Sir David, such testimonials, set against other evidence and other considerations, did not sway the Town Council. Forbes was well aware that his own scientific credentials could scarcely match Brewster's and cannily reminded the patrons that 'Scientific Celebrity' is 'far from being the principal, much less the sole, qualification'[17]. Influential voices in the Edinburgh press agreed: 'It is clear,' the *Edinburgh Weekly Journal* echoed, 'that other qualifications besides mere celebrity in any particular department of science *may be sought for, and valued in a teacher* . . .'[18].

In the Town Council's view those 'other qualifications' very probably went beyond teaching ability. The contest for the natural philosophy chair was conducted during the height of Reform agitation and elections to Parliament and burghs, and 'the spirit of party was then running very high in Edinburgh as elsewhere'[19]. Forbes was a Tory, the scion of a Tory banking family of considerable influence in the city; Brewster was a Reform Whig, recently knighted through the interest of his friend the Lord Chancellor Henry Brougham. But even the Reform Whig newspaper *The Scotsman* could not bring itself to support Brewster. First *The Scotsman* canvassed the idea of offering the chair to Sir John Herschel, then it came out strongly in favour of neither Forbes nor Brewster but of Thomas Galloway, teacher of mathematics at Sandhurst, on the grounds of his *mathematical* capabilities and his experience in teaching. It opposed Brewster on the bases of his 'great infirmities of temper', the suspicion that he would 'make the chair a sinecure, and leave its duties to some raw inferior substitute', and the belief that the great experimentalist was 'but a slender mathematician'[20]. When in January it became clear that Forbes was likely to be the victor, *The Scotsman* pinned its party colours to the mast: 'The thing', it thundered, 'is a palpable job. The old Tory banking-house was defeated in its attempt to nominate a member [of Parliament] for the city, and as a solace for the disappointment, the Town Council have surrendered the chair of Natural Philosophy into its hands!'[21].

Understandably enough, personal relations between Brewster and his former protégé were frozen for several years. A thaw was signalled late in 1836 when Forbes sent Brewster a gift of £500, almost as much as Brewster lost through Forbes' success. 'I am sure', Forbes wrote, 'that you will accept it in the Cordial Spirit in which it is offered, as an Acknowledgement of your kindness and support at a time when friends were few and difficulties harassing'[22]. Brewster was hesitant in accepting largesse from this quarter, but suggested that Forbes might render a greater service by using his influence to secure for Brewster a paid scientific consultancy to Trinity House, such as was recently arranged for Michael Faraday[23].

The loss of the Edinburgh natural philosophy chair was, as Mrs. Gordon said, 'perhaps the most severe disappointment of Brewster's life'. Because of 'extremely embarrassed' financial circumstances he was driven to apply for a situation which he knew was likely to make him unhappy and to make an active research life even more difficult than it already was. He had 'no private means, no regular profession, no remuneration from his inventions, and his greatest literary undertaking [the *Edinburgh Encyclopaedia*] having proved a complete failure in a pecuniary sense, with three sons to send out into the world, his spirits often sank at his prospects.' From the start of his working life until the Crown appointment to the St. Andrews principalship put an end to 'all these embarrassments' Brewster fought an incessant battle to meet his bills, if not to avoid utter ruin, and to buy time to do scientific research[24]. The natural philosophy chair would have brought him about £700 per annum, that is, if he could have maintained Leslie's average class-size[25], and it was a reliable source of income he desperately needed. Until 1838 Brewster's income was almost entirely derived from casual sources: small sums from editing the *Edinburgh Magazine* and the *Edinburgh Philosophical Journal* with Professor Jameson, later £100–£115 a number from Blackwood for the *Edinburgh Journal of Science*[26], a prize of 1500 francs from the French Institute in 1816, perhaps a few thousand pounds from the kaleidoscope (when he might have made £100,000 with suitable patent protection), a £300 *ex gratia* payment for nine-years' demanding service as General Secretary of the Royal Society of Edinburgh (whose duties included conducting the Society's correspondence and supervising its *Transactions*—duties he found too much for him in the last year of his tenure), a grant of £100 from the Government (raised to £300 in 1836), variable amounts from his many works of popular science, biography, edited scientific lectures and texts, and contributions to periodicals, the occasional piece of plate from scientific prizes, and whatever his services as an administrator of scientific organisations could command. And, throughout it all, from 1808 until 1830, the spectre of the *Edinburgh Encyclopaedia*, the bitter negotiations with publishers Blackwood, Constable, Murray and Tegg, the fecklessness of contributors, and the long law-suit that threatened to break him[27]. So insecure did he feel that he seriously considered Henry Brougham's offer of a living in the Church of England, Evangelical Presbyterian that he was[28]. From the 1810s until the very end of his life Brewster waged war against the iniquitous patent laws in Britain 'where inventors are more cruelly taxed than in any other part of the world'[29], laws which had robbed him of full rewards from his many inventions.

Cut off as he was by his speaking disabilities from extra-mural lecturing and from a scientific professorship Brewster could not afford to overlook any plausible source of income or to accept any situation which threatened to drain that which he had. Even the knighthood he so earnestly sought for himself and for other British men of science as a mark of the nation's esteem became a cash-consideration. When made aware that the fees might amount to £109, or £240 if he did not come to London, he was moved to decline it, and only accepted when, after remonstrances to Henry Brougham, the fees were waived[30]. And, as Jack Morrell and Arnold Thackray have shown, one of Brewster's chief interests in the British Association he did so much to found was the idea that it might lobby the Government to provide a physical laboratory and research grants for him, a hopeful expectation which bore Brewster only bitter fruit[31]. Inevitably, his dealings with scientific societies were coloured by his personal career and financial situation. For example, in an article in the *Edinburgh Encyclopaedia* he deprecated the charter of the Royal Society of Edinburgh: 'it will scarcely be believed', he wrote, 'that at such a period of liberality, they are actually prohibited from appointing a professor, lecturer or doctor of mineralogy, geology, or natural history', and strongly recommended that they should create such a paid position, almost certainly with himself in mind[32].

It was probably the severe pressure bearing upon Brewster that accounts for the most bizarre episode in his dealings with Edinburgh scientific institutions. In 1821 Brewster was closely involved with the founding of two new and separate local organisations. In the *Edinburgh Philosophical Journal* Brewster had been urging the establishment of a Scottish Society of Arts, in imitation of the London Society, for the purpose of encouraging and rewarding Scottish inventions. With Brewster as one of its leading lights the Society of Arts was almost immediately successful, enlisting the patronage of an impressive number of Scottish aristocrats[33]. In the same year Brewster consulted with Leonard Horner in planning the Edinburgh School of Arts for the scientific education of local artisans and mechanics, the first of scores of mechanics' institutes established throughout Britain by the 1830s, and Brewster was also on its first Board of Directors[34]. This too was an instant success, with over 450 students in its first year. Preparations for the second year of the School of Arts were proceeding smoothly, when, in August of 1822 Brewster embarked on a course of action which almost destroyed the enterprise. He evidently convinced himself that Horner's School was to be only a temporary institution, intended to merge with the larger, more opulent and prestigious Society of Arts[35]. Without informing Horner, who was far from sharing Brewster's conception of the relationship between the two organisations, Brewster and his colleague John Robison set about the establishment of another institution to deliver scientific lectures to Edinburgh's artisans. He fixed upon the name of the 'School of Arts' (comma) 'under the direction and patronage of the Society of Arts for Scotland'. Brewster approached Andrew Fyfe, who had given the chemistry lectures in the first year of Horner's School, and asked him to perform a similar function for the new School. Fyfe agreed, stipulating that his lectures for Brewster's School should not conflict with those for Horner's. In the event, Brewster and Robison went ahead and printed up a circular announcing their new institution, giving the same lecture times as those of the established School. So, by early September, barely a month before term was to start, the Edinburgh artisan thirsting for scientific instruction was puzzled to find that he now had a choice between two identically-named institutions, offering lectures at the same time, on the same subject, by the same lecturer.

Horner, who never liked Brewster personally and thought he had 'no capacity for great general views', was not amused[36]. Summoning a general meeting of 300 of the contributors to the School of Arts, Horner offered an analysis of Brewster's motivations:

'... very soon after the successful commencement of the School of Arts, Dr. Brewster showed an uneasiness that it was occupying so much of the public attention, as he imagined it would materially interfere with the new Society he was about to establish, and indeed avowed that it would draw away subscriptions from it. From that time he shewed great coldness and indifference to the School of Arts...'

According to Horner, Brewster then went on to try to alienate several of the lecturers and to insinuate that most damaging of allegations in the Edinburgh context, that 'the School of Arts had assumed a political character'[37]. Cockburn lent Horner strong support: 'Brewster', he said, 'being the author and master of [the Society of Arts], it was perfectly natural for him to be ambitious of making every scientific institution in the country subordinate to it'. Brewster had few allies in attendance, and he himself, while seen outside the hall shortly before, was not present. His plan was roundly defeated; his proposed School never materialised, and Brewster's name, understandably enough, was dropped from Horner's Board of Directors. In a state of high satisfaction Horner wrote to his friend Dr. John Marcet: 'Brewster's iniquities have been fully exposed... [He] has, as yet, made no attempt at a reply. He had better not, for I understand everything, and have it in my power to make him appear still more contemptible than he does already'[38].

While his Edinburgh contemporaries readily referred Brewster's behaviour to imperfections in his character, the historian is better advised to understand it against the backdrop of the pressures and constraints operating on him at that time. He was then forty-two years old, he wanted to devote his life to science and invention, and he still lacked the institutional framework and financial security so necessary to his chosen career. By 1830, when he wrote his famous review of Babbage's *Reflexions on the Decline of Science in England*, his personal situation was no better, and in many ways it was worse. One can therefore appreciate the personal dimension to Brewster's catalogue of British parsimony and Continental liberality to men of science: Cosmo, Grand Duke of Tuscany gave Galileo an annual salary of 1000 florins, exempting him 'from all professional duty' save the occasional lectures to 'sovereign princes'; His Holiness the Pope granted Galileo a further pension and Grand Duke Ferdinand an additional 100 crowns; Ferdinand I of Denmark granted Tycho Brahe 3000 crowns a year and built him an observatory; the Emperor Rudolph gave Kepler a liberal pension; Descartes received 3000 crowns from Louis XIII; Huygens won similar largesse from Colbert and the Sun King; Leibniz enjoyed the liberal patronage of the Holy Roman and Russian Emperors and he was able to leave a fortune of 60,000 ducats 'accumulated in sacks, in various kinds of specie'. And Brewster continued the catalogue of Continental munificence, recounting the precise amounts of princely pensions enjoyed by the Bernoullis, Euler, Lagrange, and Laplace[39]. 'In this enumeration', Brewster concluded, 'England holds a very subordinate place. Her liberality to Newton [knight and Master of the Mint] is the only striking instance which we have been able to record, because it is the only one in which the award of a title was combined with an adequate pecuniary reward.' The situation is 'still more distressing, when we consider the condition of our living [British] philosophers' contrasted with their Continental counterparts: the pensions to the members of the Paris Academy of Sciences, the concern of the Prussian royal family for science, even the honours bestowed by Sweden and Norway on their most distinguished chemists and geographers, the patronage given to Italian mathematicians, and on and on. '[W]e are authorized to conclude that in every nation on the continent of Europe, with the exception of Turkey, and perhaps of Spain, scientific acquirements conduct their possessors to wealth, to honours, to official dignity, and to the favour and friendship of the sovereign.' Britain had responded to this invitation to emulate by rearing up a generation of modern martyrs of science:

> 'There is not at this moment, within the British Isles, a single philosopher, however eminent have been his services, who bears the lowest title that is given to the lowest benefactor of the nation, or to the humblest servant of the crown! There is not a single philosopher who enjoys a pension, or an allowance, or a sinecure, capable of supporting him and his family in the humblest circumstances! There is not a single philosopher who enjoys the favour of his sovereign or the friendship of his ministers!'[40]

Of these new martyrs of science Brewster was, of course, the paragon, and his personal experiences provided him with the most striking materials for his portrait of the distressed state of British men of science in the early 1830s. However, nothing could be more misguided than to suggest that Brewster's complaint was animated solely by the desire for self-aggrandisement. Brewster held to an intensely moral, religious, humanitarian and nationalistic vision of the scientific enterprise, and it was only, in his view, by the proper remuneration and recognition of the labours of men of science that British science could serve God and the nation. As Paul Baxter has shown, Brewster and his fellow Scottish Evangelical men of science argued that the study of the natural world was one of the most effective ways of displaying God's providence and of instilling in mankind that sense of wonder and reverence which was the best surety for good religion and moral order[41]. Science was also the hand-maiden of civilisation itself: it released man from the chains of superstition and credulity which were the tools of tyrants and oppressors, and it laid the foundations of political liberty[42]. And Christianity and civilisation in turn supported each other, just as each was supported by science. Moreover, the cultivation of science and the arts was vital to the prosperity of the nation; as early as the 1830s Brewster was sounding the alarm about the consequences for Britain's place in the world of her niggardly treatment of inventors and men of science: 'If this goes on, John Bull may in our day order his hogshead of porter from the French capital'[43]. Finally, Brewster argued that the advance of science was the best guarantee of peace on earth. Nor was he shaken in this conviction by the contribution of science to the technology of warfare. Towards the end of his life, in the shadow of high-technology Continental war, he enunciated a Victorian version of the doctrine of 'mutual assured destruction':

> '... while art and science are thus adding to our social blessings, and are pre-eminently the instruments of peace, they have in our day been busily and successfully employed in forging the weapons of violence and destruction. Nor is this [he assured his audience] a retrograde step in civilization. By increasing the dangers, we diminish the chances, of war. In perfecting the machinery of Death, we eventually add security to Life. War may become so disastrous in its consequences, so indiscriminate in its slaughter, and so appalling in its carnage, that it will cease to be the arena of the heroic virtues; and this bloody scourge of humanity ... the master crime of nations ... will be crushed by the genius of art, and perish by the weapons itself has used'[44].

(Marvellous prose; faulty political analysis.)

We may now appreciate how Brewster's personal predicament, joined to his moral conception of the proper place of science in Christian civilisation, thrust him into the vanguard of the campaign to set British science on a more professional footing. Morrell and Thackray have demonstrated how Brewster hoped that the new British Association for the Advancement of Science, which he promoted and helped to found in 1831, would act as a vigorous scientific lobby, winning from a reluctant Government that which was due to science as a national benefactor[45]. And, if Brewster himself gained little from the Association's activities, other men of science did materially benefit. As for himself, Brewster eventually secured the knighthood he craved, a modest pension, and later the Principalships of St. Andrews and Edinburgh, and the Presidency of the Royal Society of Edinburgh which were the tokens of recognition of a life expended in under-rewarded scientific research. Honorary degrees, fellowships in scientific societies, medals and prizes he had in plenty, but until his late middle age Brewster did not have what he wanted more than anything else: the security, financial reward and leisure he reckoned was owing to a servant of the national interest. Brewster was like Moses in Sinai, leading British men of science to the Promised Land of professionalisation, but destined never to enter it himself. Still, it could have been with immense self-satisfaction that Brewster heard the Prince Consort's Presidential Address to the British Association meeting of 1859:

> '[We hope] that by the gradual diffusion of Science, and its increasing recognition as a principal part of our national education, the public in general, no less than the Legislature and the State, will more and more recognize the claims of Science to their attention; so that it may no longer require the begging-box, but speak to the State like a favoured child to its parent, sure of his parental solicitude for its welfare; that the State will recognize in Science one of its elements of strength and prosperity, to

foster which the clearest dictates of self-interest demand'[46].

It was a recognition which Brewster laboured his whole life to instil in the State, and, once instilled, it has lasted—until very recently.

When David Brewster began his working life he was a 'philosopher' and a 'man of letters'; at its end he found that he was a 'scientist'. What was involved was more than a semantic shift; it was a great transformation in cultural perceptions and social roles. The 'scientist' is paid to do specialised research the social value of which is recognised by the State and the institutions supported by the State. David Brewster's career spans this transformation; but the bulk of his life provides us with a vivid image of what it was like to be a self-employed scientific scholar.

Notes and References

I should like to thank Jack Morrell and Alison Morrison-Low for drawing my attention to several of the sources referred to here. For permission to quote from material in their keeping, I am grateful to the Trustees of the National Library of Scotland and the Library of St. Andrews University.

1. John Campbell Shairp, Peter Guthrie Tait and A. Adams-Reilly, *Life and Letters of James David Forbes, F.R.S.* (London, 1873), 59 (quoting letter from Brewster to Forbes, 11 February 1830).
2. [David Brewster], 'Decline of Science in England', *Quarterly Review*, 43 (1830), 325–326.
3. *Letters Chiefly Connected with the Affairs of Scotland, from Henry Cockburn . . . to Thomas Francis Kennedy, M.P.* (London, 1874), 137–138 (letter from Cockburn to Kennedy, 27 February, 1826).
4. *The Edinburgh Weekly Journal* (14 November 1832), 365.
5. J. B. Morrell, 'Science and Scottish University Reform: Edinburgh in 1826', *British Journal for the History of Science*, 6 (1972), 39–56.
6. Shairp *et al.*, *op. cit.* (1), 57.
7. *Ibid.*, 59.
8. Forbes assured those from whom he sought testimonials that it was with Brewster's 'perfect approval that I now stand' and that 'Brewster does not now intend to present Testimonials'. If this claim was true at the time of writing, it very soon became inaccurate. (Draft of a letter from Forbes to his prospective referees, 1 December 1832: St. Andrews University Library, Forbes Papers 1831/166.)
9. M. M. Gordon, *The Home Life of Sir David Brewster* (Edinburgh, 1869), 59, 65.
10. *Ibid.*, 94–95; *Testimonials in favour of Sir David Brewster as Candidate for the Chair of Natural Philosophy . . .* (Edinburgh, 1832), 5–7.
11. Jack Morrell and Arnold Thackray, *Gentlemen of Science: Early Years of the British Association for the Advancement of Science* (Oxford, 1981), 44 (citing letters from Brewster to Charles Babbage, February 1829, British Library, Add. Ms 37184, ff 201–203).
12. *Testimonials, op. cit.* (10), 3. In Forbes' soliciting letter (cited in note 8 above) he took care to remind his correspondents of Brewster's 'incapacity for speaking, his age, and his many personal enemies'.
13. Brewster, *op. cit.* (2), 328.
14. Gordon, *op. cit.* (9), 57.
15. *Testimonials, op. cit.* (10), 32–33.
16. *Ibid.*, 16–17, 30, 35.
17. *Testimonials in favour of James D. Forbes as Candidate for the Chair of Natural Philosophy . . .* (Edinburgh, 1832), v.
18. 'Chair of Natural Philosophy', *Edinburgh Weekly Journal* (26 December 1832); see also issues of 9 January 1833, 23 January 1833, 6 February 1833.
19. Shairp *et al.*, *op. cit.* (1), 86.
20. 'Chair of Natural Philosophy', *The Scotsman* (26 December 1832).
21. 'The Vacant Chair', *The Scotsman* (16 January 1833).
22. Forbes to Brewster, 13 November 1836; St. Andrews University Library, Forbes Papers, 1836/56.
23. Brewster to Forbes, 19 November 1836, (*ibid.*, 1836/59); see also Brewster to Forbes, 18 November 1836, (*ibid.*, 1836/58), and 21 November 1836 (*ibid.*, 1836/61). Brewster returned the money, and the suggested consultancy did not materialise.
24. Gordon *op. cit.* (9), 153, 164.
25. Calculations of the chair's salary (£22. 4. 4. from the Town Council and £30. 0. 0. from the Crown) and class fees (£638. 8. 0. from an average of 152 students) are in 'Chair of Natural Philosophy' *The Scotsman* (10 November 1832).
26. Mrs. Margaret Oliphant, *Annals of a Publishing House. William Blackwood and His Sons. Their Magazine and Friends*, 3 vols (Edinburgh and London, 1897–98) II, 8–9; see also W. H. Brock, 'Brewster as a Scientific Journalist', in this volume.
27. See, for example, Gordon, *op. cit.* (9), 95–97, 164; *Transactions of the Royal Society of Edinburgh*, 11 (1831), 509; 'Minutes of Statutory and Special General Meetings [of the Royal Society of Edinburgh] . . . 1812 to 1926', (Minute of 5 January 1829); Morrell and Thackray, *op. cit.* (11), 41 no. 31, 43–44. In his letter to Forbes on 19 November 1836 (see note 23 above) Brewster estimated his potential liability for the debts of the *Edinburgh Encyclopaedia* at £1500.
28. Gordon, *op. cit.* (9), 154. The Anglican living proposed by Brougham would have brought Brewster £700 *per annum*. Brewster consulted extensively with the Bishop of Cloyne about this idea; the Bishop, according to Brewster, 'warmly approved of it and offered me private ordination'. But Brewster's plan was put to one side by affairs involving his wife's family's estate of Belleville, and, when debts began to accumulate there as well, it was too late to take up the ecclesiastical venture: as Brewster wrote to Forbes, 'the Lord Chancellor [Brougham] was out of office, the Bishop of Cloyne dead, and ordination at my age impossible'. Brewster to Forbes, 19 November 1836 (see note 23 above).
29. David Brewster, 'Opening Address, 4 December 1865', *Proceedings of the Royal Society of Edinburgh*, 5 (1866), 459.
30. Letter from Brewster to Babbage, 17 November 1831, British Museum, Add. Ms 37186, ff 150–151; Gordon, *op. cit.* (9), 150.
31. Morrell and Thackray, *op. cit.* (11), esp. 299, 551.
32. [David Brewster], 'Societies', *Edinburgh Encyclopaedia*, 20 vols (Edinburgh, 1830), XVIII, 268; see also Brewster, *op. cit.* (2), 321.
33. [David Brewster], 'History of Mechanical Inventions and Processes in the Useful Arts', *Edinburgh Philosophical Journal*, 5 (1821), 115.
34. 'Some Account of the School of Arts of Edinburgh', *Edinburgh Philosophical Journal*, 11 (1824), 203–205; *First Report of the Directors of the School of Arts of Edinburgh* (Edinburgh, 1822).
35. This account (and the quoted material below) is from a report of the General Meeting of Contributors to the School of Arts, 3 September 1822, in *The Scotsman* (7 September 1822).
36. Letter from Leonard Horner to Dr. John Marcet, 10 April 1821: National Library of Scotland Ms 9818, ff 88–89.
37. For another episode illustrating Horner's sensitivity to political sectarianism in Edinburgh cultural institutions:

Steven Shapin, '"Nibbling at the Teats of Science": Edinburgh and the Diffusion of Science in the 1830s', in Ian Inkster and Jack Morrell (eds.), *Metropolis and Province: Science in British Culture, 1780–1850* (London, 1983), 151–178.
38. Letter from Horner to Marcet, 14 September 1822, NLS Ms 9818, ff 103–104; also Katharine M. Lyell, *Memoir of Leonard Horner,* 2 vols (London, 1890), I, 204.
39. Brewster, *op. cit*. (2), 309–315.
40. *Ibid.*, 320.
41. Paul Baxter, 'Brewster, Evangelism and the Disruption of the Church of Scotland', in this volume.
42. Gordon, *op. cit.* (9), ch. 18; Richard S. Westfall, 'Introduction', to Brewster, *Memoirs of the Life, Writings, and Discoveries of Isaac Newton* (Reprinted from the Edinburgh edition of 1855: New York, 1965), ix–xlv, esp. xv–xxii.
43. Brewster to Henry Brougham, 10 March 1828, Brougham Papers, University College London, quoted in Morrell and Thackray, *op. cit.* (11), 41.
44. Brewster, *op. cit.* (29), 466.
45. Morrell and Thackray, *op. cit.* (11), esp. chs. 2 and 6.
46. Prince Albert, '[Presidential Address] at the Meeting of the British Association for the Advancement of Science. [Held at Aberdeen, September 14th, 1859]', in *The Principal Speeches and Addresses of His Royal Highness the Prince Consort* (London, 1862), 227.

Brewster and the early British Association for the Advancement of Science

J. B. MORRELL

A bi-centennial celebration can fill one with gloom, especially if one is recruited to dilate upon a savant who is, historically speaking, rather tedious and unproblematic. In the case of Sir D.B., as his friends called him, we have a welcome exception and indeed a glorious opportunity: Brewster was never dull; and this Symposium has revealed that his extraordinary career poses many interesting historical problems. Perhaps one aspect of his work has not been sufficiently stressed today, namely, that at his best Brewster is one of the great writers of English prose, particularly in that characteristically Victorian genre, the long essay review. In his zealous and obsessed moods, when his pen could hardly keep pace with his mind, or with grammar, his writing was fiery and persuasive. Here is one of my favourite gobbets from Brewster, his denunciation in 1830 of the iniquities of the patent law:

> 'a system of vicious and fraudulent legislation, which ... places the most exalted officers of the state in the position of a legalized banditi, who stab the inventor through the folds of an act of parliament, and rifle him in the presence of the Lord Chief Justice of England'[1].

For the historian such splendid prose is a pleasure to read but it presents problems of handling. With respect to the British Association the problem is acute because by 1851 Brewster had put into print three retrospective accounts of its origins and early history; the first in the *Edinburgh Review* in 1835, the second in the *North British Review* in 1850, and the third in his Presidential address to the Association also in 1850[2]. At the factual level, these accounts can be misleading, sometimes seriously so. For instance, in the 1850 article Brewster made much of the views on the decline of science allegedly proclaimed by William Vernon Harcourt in his inaugural speech to the Association in September 1831. This 1850 account of Harcourt's speech was followed by Mrs. Gordon in her biography of Brewster[3]. More recently Derek Orange rightly characterised Brewster's 1850 version as odd[4]. This oddity is explained if we turn to contemporary newspaper accounts of Harcourt's speech and contemporary correspondence, from which it is clear that Brewster's memory deceived him. Nineteen years after the event, Brewster cited a *letter* Harcourt wrote in late November 1831 to Lord Milton, a copy of which Harcourt sent to Brewster, as being what Harcourt *said* in his speech of September 1831. In our book on the early Association, to which this paper is deeply indebted for empirical material and interpretative framework, Arnold Thackray and I were particularly careful when exploiting Brewster's retrospective public accounts: we found them very stimulating for his polemical perceptions and acute insights, but whenever possible they were checked against published and unpublished contemporary documents[5].

One of Brewster's most imaginative and successful experiments was to instigate the British Association for the Advancement of Science. In this paper I want to look at the changing relations of Brewster to the Association, beginning with his proposing the formation of an association of nobility, clergy, gentry, and philosophers to remedy the depressed state of British science. He led the northern lights at the inaugural York meeting of the Association in 1831, and by the second meeting at Oxford in June 1832 he was as enchanted as ever with it: he had successfully drawn attention to the importance of establishing a Council, his report on optics confirmed his high intellectual position in the Association, he was one of two Vice-presidents at the 1832 meeting, and along with three other religious dissenters he had received during the meeting the rare accolade of an honorary doctoral degree from the University of Oxford. Then, I shall argue, the Association for which Brewster had campaigned so energetically began to be used against

Fig. 3. Presidential Banner of the British Association for the Advancement of Science for the 1850 Edinburgh meeting when Brewster was President. (*British Association for the Advancement of Science*).

him. In early 1833 his protégé, James David Forbes, drew considerably on his own Association contacts to defeat Brewster in a contest for the chair of natural philosophy at the University of Edinburgh. At the Cambridge meeting in 1833 the advocates of Fresnel's wave theory of light, mainly drawn from Trinity College, Cambridge, and Trinity College, Dublin, began to promulgate that theory in Section A of the Association, thus challenging Brewster's previous dominance there. By 1834 it was patently clear to Brewster, a Vice-president that year, that the Cambridge invasion of the Association meant that it would not be the reforming body he envisaged. In 1835 he vented his spleen in the *Edinburgh Review*, an act of apparent disloyalty which provoked Thomas Romney Robinson to refer publicly at the 1835 meeting to Brewster as one who wielded 'the concealed dagger of a lurking assassin'[6]. There was indeed rich irony in Brewster's relation to the early Association: though at its centre until 1832, he had been edged to its periphery only two years later.

I

By the late 1820s, when he was almost 50 years old, Brewster had earned a living for thirty years as an editor, author, encyclopaedist, inventor, and consultant. Though he acquired in 1829 a government pension of £100 per year, Brewster felt he lacked regular recognition commensurate with his talents. From the early 1820s he had been irritably dissatisfied with the treatment of scientific men by government; and in the late 1820s he took every opportunity, *via* his periodical, the *Edinburgh Journal of Science*, to press for increased state patronage of science and scientific men. If such concerns were generalisations from his own case, the deaths of William Hyde Wollaston, Thomas Young, and Humphry Davy, in just six months between December 1828 and May 1829, staggered him; they also confirmed his view that an era in British science had ended and that the time was ripe for the scientific millennium to be at last ushered in.

With respect to organised science in the Scottish metropolis, Brewster was irritated in the late 1820s with the Royal Society of Edinburgh of which he was Secretary for six years. He thought its elections were mismanaged. He deplored the refusal of the Society to elect as a Fellow his young protégé, James David Forbes; he resented the difficulties concerning the election of another protégé, the chemist James Finlay Weir Johnston. He was outraged by the reluctance of the Society to pay the legal costs generated by his attempts as Secretary to extract the Society's property from the University museum where Robert Jameson zealously guarded his geological booty. And Brewster reprobated the annual rent of £260 for its apartments which the Society paid to government[7].

By 1829, then, Brewster's restlessness made him receptive to Charles Babbage's article about the 1828 Berlin meeting of the Gesellschaft Deutscher Naturforscher und Ärzte which from 1822 had met annually in a different German city[8]. Like Babbage, he was inspired by the German congress of savants and brooded on the idea of a European scientific academy. Again like Babbage he was obsessed with the question of the decline of science, for him a heart-breaking subject. It was not surprising that in early 1830 when Babbage had begun writing his book, *Reflections on the Decline of Science in England*, he turned to Brewster for ammunition. In reply Brewster tried without success to interest Babbage in organising 'an association for the purpose of protecting and promoting the *secular* interests of science'[9]. When Babbage's book was published in late spring 1830, Brewster returned in private to the notion that an association should be set up to revive science and to act as a pressure group on government. Publicly he reviewed Babbage's book in the October 1830 issue of the *Quarterly Review*: in a long article on the general decline of science, he proposed that an association of nobility, clergy, gentry and philosophers should be formed to remedy the depressed state of science. Brewster remained convinced that ultimately government alone was capable of patronising science properly, so that for him his proposed association was a necessary preliminary vehicle which would arouse government to its responsibilities. Brewster's *Quarterly Review* article also carried some cavalier remarks about English universities and roused the wrath of both William Whewell and George Biddell Airy. By late 1830 Brewster's zealous excitability had earned for him the hostility of the two leading Cambridge mathematical physicists[10].

Early in 1831 Brewster received for his journal an article from Johnston about the 1830 Hamburgh meeting of the German naturalists. This account confirmed Brewster's opinion that it was desirable to have an equivalent British institution[11]. On 21 February 1831 he wrote to Babbage, specifying York and July/August as the time and place of the first meeting; and with crass naïvety he suggested for President, John Herschel, who had just been defeated by the Duke of Sussex in a contested election for the Presidency of the Royal Society of London. Two days later he optimistically informed John Phillips, secretary of the Yorkshire Philosophical Society, that arrangements for the meeting to be held at York were 'in progress'[12]. Brewster's views about the date, the size, the finances, and the aims of the Association, were not to be implemented. But through these two letters of February 1831 he tolled the bell that called together two groups which were to be of central importance at the first meeting; the Edinburgh savants (himself, Forbes, Johnston, John Robison and Sir Thomas Makdougall Brisbane); and the Castor and Pollux of York science (Harcourt and John Phillips).

Proposing is one thing, disposing another. From February 1831 until late September 1831 when the Association met in York, Brewster maintained his roles as propagandist and projector but declined that of organiser. Encouraged by the enthusiastic response from York, in April 1831 Brewster gave notice of the meeting in his own journal, making it public that in his view the Association was to be modelled on the German gathering, was to meet in July at York, and that Robison (the Secretary of the Royal Society of Edinburgh) was to be interim secretary. Having passed the difficult organising of a novel event to Robison, Brewster become so absorbed with his spectroscopic researches that, as Robison complained, he became a member of an anti-corresponding society. In July 1831 Brewster emerged by publishing in his journal his draft plan of nine regulations for the constitution of the proposed body, borrowing heavily from the German model. With characteristic unpredictability, he called publicly on Robison, Johnston, and especially the Yorkshire Philosophical Society to be prepared in September with a code of laws. Agitation of this kind ensured that the proposed meeting was seen by Brewster's opponents as heavily identified with him. That was the chief reason why the constellation of savants from Trinity College, Cambridge, did not attend the York meeting: to them, going to York was equivalent to rallying round Brewster's standards, that is, subscribing to his views on the decline of science. In fact, from early July 1831 the lead in organising the meeting and framing its constitution was taken not by Brewster but by Harcourt, a masterly politician who could be all things to all men without seeming disingenuous: a man capable of seeming a declinist to Brewster yet able to win Whewell to the cause of the Association. At the York assembly Harcourt's dominance was such that even Brewster acknowledged there and then that Harcourt had been the soul of the meeting. As Harcourt himself was to encapsulate the matter in 1853, it was Brewster who proposed that a craft be built for the united crew of British science, but it was Harcourt who manned the ship, constructed her charts, and piloted the vessel[13].

At the 1831 meeting one of Harcourt's master strokes was to name Brewster and Whewell as Vice-presidents for the 1832 meeting at Oxford. Between these two meetings, Brewster both promoted and embarrassed the fledgling Association. He recruited a few Scottish members, though nothing like as many as Forbes who induced forty men to join. In April 1832 Brewster told Harcourt that the Association ought to be governed by a permanent Council, modelled on the French Institute, through which of course he hoped to lobby for his pet scheme of a state-supported physical laboratory, to be established especially for him. Brewster also realised the advantages of having a small decision-making body to act between meetings. The Association, characterised by Brewster as Frankenstein's monster, needed a flywheel. This notion was highly agreeable to those from Cambridge, Oxford and London who wished to increase their control of the *parvenu* Association. At the 1832 meeting, constitutional revision was therefore smoothly and quietly accomplished when the Council was established to run the Association 51 weeks of the year. Ironically, the Council which Brewster proposed was one means by which the Cambridge/London/Oxford axis, conspicuously absent in 1831, took over the Association from 1832.

Brewster also caused Harcourt some embarrassment in late 1831 and early 1832. Though Harcourt *privately* maintained that British science was stagnating and *privately* favoured direct state patronage of individuals, *publicly* he adopted a position which put him in neither the declinist nor the anti-declinist camp. In Harcourt's view the issue of state patronage was prospectively divisive. That is why he tried to mute Johnston's account of the York meeting published in Brewster's *Edinburgh Journal of Science* in January 1832. Brewster thought Johnston had written in a good spirit, but Harcourt disagreed. He had just managed to recruit Whewell to the Association and he felt there was a real danger that Johnston would alienate the Cantabs who were so essential to Harcourt's plans, for instance, as reporters on the state of science. Johnston refused to be muzzled by Harcourt: he referred pointedly to jealousy of the Association by leading savants; and he argued that the Association should try to arrest the decline of science in Britain. Moreover Johnston accused Lord Milton, the Association's first President, of being out of touch with general opinion when in his speech at York the nobleman had reprobated the *direct* encouragement of science by government[15].

Despite Brewster's continued obsession with the debate on the decline of science, and his support for Johnston's views, he was a central figure at the 1832 meeting at Oxford. He produced his report on optics, and along with three other religious dissenters (Robert Brown, Michael Faraday and John Dalton) he received an honorary Doctor of Civil Law degree from the University of Oxford[16]. This was a highly political act, just after the third Reform Bill had become law: it allowed the University to make a timely answer to accusations of bigotry and reaction, though at the cost of seeming to make a move which might lead to the abolition of religious tests at Oxford. The Association boasted about its absence of religious barriers, and thus presented a model of benign toleration to Oxford, the renowned citadel of the established church. No doubt Brewster was delighted with his Oxford degree as a personal accolade, just as he welcomed it as another victory for dissent, for whiggery, and for reform.

II

As a loyal Whig Brewster was knighted in late 1831, an honour which did little to improve his finances. Lacking private means and a regular profession, in spring and summer 1832 he turned to his patron Lord Brougham for either a better pension or an English church living. In Brewster's view the latter would permit scientific research better than writing or inventing. After negotiations involving the Archbishops of York and Canterbury, the astronomer John Brinkley (then Bishop of Cloyne) and Harcourt, Brewster finally rejected this particular way out of his financial difficulties[17].

In late 1832 the chair of natural philosophy at the University of Edinburgh became vacant through the death of Sir John Leslie, Though he had been a stern critic of Scottish universities, Brewster became a candidate because he was desperate for a job. To his astonishment Brewster discovered that his own candidature was opposed by James David Forbes, a well-off son of the late Sir William Forbes, the leading Edinburgh banker. Forbes was an ambitious and ingratiating man, much younger than Brewster, greatly his inferior in scientific reputation, and of course heavily indebted to Brewster, his chief Scottish scientific patron. Nothing daunted, Forbes launched his canvas with the Town Council of Edinburgh, the electors to the chair. He desperately needed evidence *via* testimonials that he was a national figure. It was here that Forbes exploited his contacts in the Association and the work he had done for it, such as his 1832 report on meteorology. Testimonials from Harcourt and Whewell, for instance, were in the Forbes's hands before Brewster had even applied to them.

By the end of 1832 Forbes had printed 59 testimonials to himself. No fewer than 14 of these referred glowingly to his work for the Association. From York, Harcourt stressed that Forbes was a valued friend, while Phillips went so far as to speak fulsomely on behalf of the Association about Forbes. Phillips subsequently regretted this rash assertion: fearful of conflict between Brewster, the Scottish instigator of the Association, and Forbes, its chief Scottish activist, he privately hoped Brewster would be appointed because that would be less dangerous to the Association's peace. For his part, in response to Brewster's plea, Harcourt stressed to the Lord Provost of Edinburgh that at the 1831 meeting Brewster had displayed considerable power as a public speaker, the Forbes' camp having rumoured that Brewster was incapable of lecturing because of nervousness. Having given a testimonial to Forbes, Harcourt was clearly trying to disable Forbes' use of the Association as an electioneering resource. Harcourt's action was subsequently followed by Phillips, Johnston and Robison, each of whom testified publicly to Brewster's prowess in *extempore* exposition. But the damage had been done. Partly through his Association contacts Forbes amassed about seventy testimonials. Another candidate, Thomas Galloway, too busy teaching at Sandhurst to attend the meetings of the Association, produced only eleven, mainly from Scottish professors, none of whom was active in the Association[18].

In the event, the Town Council elected Forbes to the chair primarily because of personal contacts between its members and the Forbes' banking house and because the Tory Town Council supported the Tory Forbes against the Whig Brewster. Yet the Town Council's wishes were legitimated by the testimonials of English philosophers, especially Cambridge and Oxford academics who were impressed by Forbes' work for the Association. For Brewster there was bitter irony in the way in which Forbes had accomplished what Brewster called 'the most scandalous job that the history of science records'[19]. After all Forbes had exploited the Association of which Brewster, first his mentor and then his opponent, was the principal instigator.

As if this were not enough, by summer 1833 at the Cambridge meeting of the Association, Brewster found himself involved in a fierce battle about Fresnel's wave theory of light. The location of the controversy was Section A, devoted to mathematical and physical sciences, a section which occupied the commanding pinnacle of the Association's hierarchy of sciences. For the Cambridge faction and its Dublin allies, the wave theory exemplified the new mathematical physics. These Cambridge and Dublin mathematical physicists sought to promote the wave theory and they found in Section A a leading instrument for propagating their views. On the other hand, Brewster, the instigator of the Association who resented the Cambridge invasion of it, found Section A an equally natural vehicle for objecting both to a new style of enquiry which threatened his position and to what he regarded as the unwarranted dogmatism of Fresnel's Cambridge supporters. At the 1832 meeting in his report on optics Brewster expressed his doubts about the wave theory. He acknowledged its power and beauty, but felt it could not cope with absorption spectra. His own recent work on nitrogen dioxide, which had almost blinded him, showed that its absorption spectrum contained more than a thousand *specific* dark lines. For Brewster, a wave theory implied the physical existence of an ether to carry the waves. He could not conceive how an ether, modified by the particles of nitrogen dioxide which contained it, could *selectively* absorb over a thousand *specific* waves of light, while transmitting the rest.

Between the 1832 and 1833 meetings the supporters of the wave theory moved into the attack, through the work of William Rowan Hamilton and Humphrey Lloyd in Dublin, of Airy in Cambridge, and of Baden Powell in Oxford. When the Association met at Cambridge in 1833, the Cambridge/Dublin coterie seized its opportunity. In his address as local secretary, Whewell stressed that the wave theory, with its correlating and predictive powers, belonged on the same level as Newton's law of gravity. Though Whewell did not mention Brewster by name, he did acknowledge that the wave theory could not begin to explain absorption spectra. A few days later, Section A brought together most of the leading protagonists: Hamilton, Lloyd, John Herschel and Airy on the one hand, *versus* Brewster, Richard Potter and John Barton on the other. In a telling discussion of absorption spectra, Herschel argued that ether particles *could* be connected with the molecules of substances to form complex vibrating systems which would resonate only to particular wavelengths. This analogy with sound, which Herschel illustrated experimentally on the spot, forced Brewster to acknowledge the problems of the emission theory of light and to admit that the facts of absorption were not incompatible with the wave theory. Very craftily Whewell and his allies commissioned a report by Lloyd on physical optics for the 1834 meeting. I suspect that Lloyd's report was deliberately intended to replace Brewster's own 1832 report on optics and it ensured that from 1834 the wave theory became the new orthodoxy of Section A. To call for and obtain a second report on a topic by a different author, within just two years of the first report, was unique in the early annals of the Association; and it doubtless added insult to the injury inflicted on Brewster.

These polemics in Section A, and their successors, showed that at the extremes contrary career investments were at stake, Brewster's in experimental optics and Whewell's in mathematical physics. Each party was strong where the other was weak: Brewster was no expert in analytical mathematical methods; Whewell had little talent for sustained delicate experimentation. Brewster resented the polemical, arrogant and partisan idea that the undulatory theory was on a par with Newton's theory of gravitation; and he deplored the rashness with which Whewell asserted the physical existence of an ether, which for Brewster was both unobservable and unimaginable. Thus the optical disputes in Section A revealed animosities of methodology, of person, of career, and of style[20].

By 1834 it had become clear to Brewster that the Association, his own brain-child, was relatively indifferent to the question of direct national provision for men of science. He was disappointed that it had neglected or renounced key issues such as government research posts, scientific Members of Parliament, pensions for scientists, honours for scientists, never mind reform of the patent laws. This emasculation of his original aims he ascribed mainly to the Cambridge invasion of the Association, an explanation which Arnold Thackray and I think is essentially correct but incomplete.

Let Brewster himself have the last words about the early British Association. At the 1842 meeting he declared publicly and ominously that learned societies, meaning the Royal Society of London and the Association, were subject to 'the incubus of the undulatory theory'[21]. He thus revived the metaphor he had used in his 1835 *Edinburgh Review* article to denigrate the Cambridge coterie. In that article he had complained that there was 'an incubus pressing on the vitals of the Association with its livid weight ... a congestion somewhere near its heart, impeding its respiration, and disturbing its most vital functions. Is it political, ecclesiastical, or personal, or is it all of them combined?'[22]. Brewster's use of the same metaphor, that of the incubus, to describe both the Cambridge faction and the undulatory theory shows how deeply he felt that he had become the odd man out, and had lost intellectual and organisational control of the Association he had so enthusiastically initiated.

Notes and References
For permission to cite from manuscripts in their care I am grateful to the Trustees of the British Library, the British Association for the Advancement of Science, and the Library of University College, London.

1. [David Brewster], 'Decline of Science in England', *Quarterly Review*, 43 (1830), 333.
2. 'The British Scientific Association', *Edinburgh Review*, 60 (1835), 363–393; 'British Association for the Advancement of Science', *North British Review*, 14 (1850), 235–287; 'Presidential Address', *Report of the British Association for the Advancement of Science ... 1850* (London, 1851), xxxi–xliv.
3. M. M. Gordon, *The Home Life of Sir David Brewster* (2nd edition, Edinburgh, 1870), 147–148.
4. A. D. Orange, 'The Origins of the British Association for the Advancement of Science', *British Journal for the History of Science*, 6 (1972), 152–176, 173.
5. J. Morrell and A. Thackray, *Gentlemen of Science: Early Years of the British Association for the Advancement of Science* (Oxford, 1981), 142 for details of Brewster's inadvertence.
6. Robinson, speech at the Dublin meeting, 15 August 1835, *Athenaeum*, 8 (1835), 642.
7. Morrell and Thackray, *op. cit.* (5), 43–44.
8. C. Babbage, 'Great Congress of Philosophers at Berlin', *Edinburgh Journal of Science*, 10 (1829), 225–234.
9. C. Babbage, *Reflections on the Decline of Science in England, and on some of its causes* (London, 1830); Brewster to Babbage, 24 February 1830, British Library, Add. Ms. 37185, f 72.
10. Brewster, *op. cit.* (1), 341; Morrell and Thackray, *op. cit.* (5), 51–52.
11. Johnston, 'Account of the Meeting of Naturalists at Hamburgh', *Edinburgh Journal of Science*, 4 (1831), 189–244.
12. Brewster to Babbage, 21 February 1831, British Library, Add. Ms 37185, ff 481–482; Brewster to Phillips, 23 February 1831, Foundation volume, British Association archives, Bodleian Library, Oxford.
13. Morrell and Thackray, *op. cit.* (5), 58–94, give full details of the respective roles assumed by Brewster and Harcourt before and at the 1831 meeting.
14. On the creation of the Council, see Morrell and Thackray, *op. cit.* (5), 298–302.
15. Johnston, 'Account of the Scientific meeting at York', *Edinburgh Journal of Science*, 6 (1832), 1–32; Morrell and Thackray, *op. cit.* (5), 143–144.
16. David Brewster, 'Report on the Recent Progress of Optics', *Report of the British Association for the Advancement of Science ... 1832* (London, 1833), 308–322; Morrell and Thackray, *op. cit.* (5), 232, 390–391.
17. Gordon, *op. cit.* (3), 157; Brewster to Babbage, 26 March, 8 April 1832, British Library, Add. Ms 37186, ff 297–298, 321–323; Brewster to Brougham, 9 May, 28 May, 30 August 1832, Brougham Papers, 15728, 26616, 15730, The Library, University College, London.
18. *Testimonials in favour of David Brewster as a candidate for the chair of natural philosophy in the University of Edinburgh* (Edinburgh, 1832); *Testimonials in favour of Thomas Galloway ...* (Edinburgh, 1832); *Testimonials in favour of James D. Forbes ...* (Edinburgh, 1832); Morrell and Thackray, *op. cit.* (5), 431–3. I am preparing a detailed study of this fascinating election.
19. Brewster to Babbage, 3 February 1833, British Library, Add. Ms 37187, ff 408–411.
20. For the disputes on the wave theory see: G. N. Cantor, 'The Reception of the Wave Theory of Light in Britain: a Case Study Illustrating the Role of Methodology in Scientific Debate', *Historical Studies in the Physical Sciences*, 6 (1975), 109–132; T. L. Hankins, *Sir William Rowan Hamilton* (Baltimore and London, 1980), 88–95, 129–171; Morrell and Thackray, *op. cit.* (5), 466–472.
21. *Literary Gazette* (1842), 534.
22. [David Brewster], 'The British Scientific Association', *Edinburgh Review*, 60 (1835), 392.

Brewster and the Reform of the Scottish Universities

ROBERT ANDERSON

Although Brewster had a long life and was a prolific writer, and although he held university principalships for thirty years, he said remarkably little about university reform. He did have views on it, which I shall attempt to piece together, but I shall suggest that his importance was partly as a symbol of reform for other people as well as being an actor in the story.

As a scientist and a leading propagandist for science, Brewster was obviously interested in introducing more science into the universities, but that aim could take different forms, and needs to be defined more closely. Besides, university reform in Scotland in the early nineteenth century was not only, or even mainly, about what should be taught, or about different concepts of what a university should be, but about the universities as public institutions—their system of government, abuses in their administration, their relations with Church, state and community. The first wave of reform feeling, which resulted in the appointment of a Royal Commission of visitation in 1826, reflected discontent with the universities as unreformed corporate bodies, which—it was alleged—had become too enclosed and inward-looking, too clerical and out of touch with the demands of the age, inefficiently managed, perhaps even corrupt. The wind of reform needed to blow through them, and the solution was seen in a form of university government which would take decisions out of the hands of the professors alone and be more responsive to public opinion. The recommendations of the Royal Commission were along these lines, but it was not until 1836 that a Whig government introduced a bill to carry them out, a bill which had to be withdrawn in the face of opposition orchestrated by the Church and by conservative university interests. This was the context, one of frustrated reform, of Brewster's appointment in 1838 by the Melbourne government to be Principal of the United College at St. Andrews. (The reforming legislation had to wait until 1858.)

St. Andrews was by far the smallest of the Scottish universities, with twelve professors and at this time about 150 students. Divided into two colleges—United College was the non-theological part—it had something of an Oxford or Barchester atmosphere, and retained some of the nepotistic flavour of eighteenth-century Scotland. There was work therefore for a reforming hand. Brewster's appointment as a layman was itself an innovation, for Scottish university principals were normally clergymen, and the income was so small that the office was often combined with a chair or a city living. Principals were not expected to be particularly dynamic figures. The original motive for Brewster's appointment may have been to provide a distinguished scientist with an income, but he arrived at St. Andrews with reforming intentions. The result was that he was soon on the worst of terms with the rest of the College and the University. Normal business relations broke down, and a series of disputes led to the appointment of a special Royal Commission on St. Andrews in 1840, whose report, published in 1845, provides the historian with a good deal of material, including oral evidence by Brewster.

The most striking thing about this report is the violence of the language used on both sides, especially by Brewster himself in his main accusation, which was that the professors were mismanaging the finances and administering the bursary funds in a corrupt manner. Brewster certainly seems to have lacked the suavity and diplomatic skills required by any successful administrator, and of course it was one thing for reformers to make general remarks about the unreformed state of the universities, quite another to make specific charges face to face in a small society of men with whom Brewster was expected to have daily contact. In the event the Royal Commission refuted

Fig. 4. St. Salvator's College and Chapel, St. Andrews, 1842. Calotype by either Dr. John or Robert Adamson. *(Royal Scottish Museum)*.

most of Brewster's charges, and reprimanded him severely for his conduct. The details of the disputes seem trivial today, but they make sense if we remember the political and especially the religious context, for Brewster was an Evangelical in a stronghold of the Moderate party. Many of his charges of financial mismanagement were in fact taken over from the Evangelical leader Thomas Chalmers who, as professor of moral philosophy at St. Andrews in 1823–28, had publicized them at the time of the earlier Royal Commission.

This link with Chalmers can also be seen in Brewster's more positive work at St. Andrews, the encouragement of science. Chalmers, along with the professor of natural philosophy, Thomas Jackson, had wanted the University to extend its teaching to cover natural history and modern languages, in order to 'make us stand complete as a School of General Education'. This would:

> 'put us in a state of special adaptation to one important class of society, that is, to parents of higher rank, who having no professional object in the education of their sons, but merely that they should be instructed in all those branches of scholarship which enter into the finished education of a gentleman, would feel it a recommendation of St. Andrews, that we are relieved from the admixture of Law and Medical Students—a recommendation quite akin to what is often stated as peculiar to our city, the very great retirement of the place.'

Since St. Andrews could not compete with the big city universities, Chalmers argued, it should exploit its natural advantages to become—in modern terminology—a liberal arts college. Jackson added a somewhat ingenuous argument for better buildings and higher salaries:

> 'As it is . . . my most anxious wish to see this University more resorted to by the young Aristocracy of our country, I could wish, in furtherance of that object, to see a little more temptation thrown in their way. Every thing offensive to a tasteful eye, in our external and internal condition, should be removed. Their Professors should be enabled to receive them with a hospitality suitable to the rank of the one party, and the liberal station and parental care and influence of the other'[1].

Brewster regretted in the same way that the corrupt state of the Scottish universities, and their failure to offer a complete course of liberal education, had driven away 'the sons of the nobility and gentry of the land'[2], and to the 1840 Royal Commission he expounded a programme designed to attract them back[3]. Like Chalmers he saw chemistry, natural history and modern languages as the desiderata. Modern languages were not to gain their place in the universities until the 1890s, but Brewster was more successful on the purely scientific front. He succeeded, in 1840, in using an existing endowment to found a chair of chemistry. He then proposed to turn the chair of medicine, which had fallen vacant, into a chair of natural history. Here he ran into opposition, for although there was no medical school at St. Andrews the existence of the chair was the basis of the University's very substantial income from awarding medical degrees *in absentia*—selling degrees, critics alleged, though the system had been tightened up in recent years. Brewster lost that battle, and the medical chair survived, but in 1850 he succeeded in converting the equally effete chair of civil history into one of natural history. In the early years, too, there was a course in civil engineering given jointly by several professors, and Brewster himself lectured regularly on scientific subjects (this had been a condition of his appointment)[4]. Thus despite Brewster's personal difficulties, and despite the questionable suitability of St. Andrews as a centre of science and technology, his achievements there were by no means negligible.

Brewster's main opponent was George Cook, professor of moral philosophy, leader of the Moderate party in the Church, and member of an entrenched St. Andrews academic dynasty. The St. Andrews disputes were at their height in 1843 when the Disruption occurred, and the latter had a direct impact when the students elected Chalmers as Rector: this was illegal since at St. Andrews the rectorship was confined to four senior figures within the University. When the Senate disciplined the student leaders involved, Brewster took their side. The Royal Commission had to adjudicate on this dispute, and inevitably found against Brewster. In view of all this, it is hardly surprising that the Senate, along with the Presbytery, seized the opportunity of attempting to depose Brewster from the principalship because he had joined the Free Church. This attempt failed, and affairs at St. Andrews eventually calmed down, especially after Cook's death in 1845.

This brings us to Brewster's symbolic period, as a hero of the Free Church and a victim, or near-victim, of persecution. The question of 'university tests' was a major political issue for ten years. All university professors were required by law to subscribe to the Confession of Faith on their appointment; that is, to declare their loyalty to the Church of Scotland, which after 1843 was no longer the national church, but the church of a minority. The Brewster case established that men already in office could not be deposed, but the Church was able to bar new professors who did not belong to the Establishment, and if this continued the universities were in danger of declining into sectarian institutions. Attempts to change the law failed until 1853, and the eventual success owed much to Edinburgh Town Council. That body, as is well known, was the 'patron' of Edinburgh University until 1858, and appointed most of the professors. It was also, after 1843, a stronghold of Free Church and Dissenting opinion, and it deliberately challenged the law by appointing Free Churchmen to chairs. It was also embroiled in longstanding disputes with the Senate of the University, which ended in defeat when the Universities Act of 1858 stripped the Town Council of its powers.

That Act also provided that all university principals might be laymen. In 1859 Principal Lee of Edinburgh died, and by a curious overlap it fell to Edinburgh Town Council to make a last appointment. They chose Brewster, and very unusually for Town Council appointments there was no contest. The Established Church candidate seems to have been John Cook, a St. Andrews clergyman who was the nephew of George Cook, but his candidature was not pressed to a vote. In appointing Brewster, I would argue, the Town Council were not thinking about how the University should be run, still less of any pro-scientific reforms in its teaching, but were simply honouring a great man, who was an eminent scientist, but also a figure in the Liberal pantheon, a representative of an older generation which had borne the heat of the day. His name was linked with that of Brougham, who had been elected as the University's first Chancellor, as 'friends of liberty and friends of science'. 'No doubt they were both full of years and honours', said Brewster's proposer, 'but the additional honours they proposed to give were the proper compliments to mature age, as well as to talent and genius'[5]. This view of Brewster's venerability was also found in Macaulay's set-piece parliamentary speech on university tests in 1845, when he derided that petty body, the Presbytery of St. Andrews, for its persecution of a figure of European reputation[6]. It was perhaps Brewster's own view too. His book *The Martyrs of Science* was published in 1841, when he was locked in combat with the obscurantist clergy of St. Andrews. Its message—found elsewhere in his writings—was one of the slow but constant progress of rationalism, the emancipation of truth from clerical domination and spiritual despotism. This was a standard Liberal-Protestant interpretation of European intellectual history, but it is not far-fetched to suppose that Brewster saw a personal application; at any rate, his inaugural address to the Edinburgh students in 1859 began by celebrating the abolition of tests and the opening of the office of Principal to laymen as triumphs of secularisation[7].

The Town Council probably saw it as a suitable parting gift to the Edinburgh professors to lumber them with a notoriously quarrelsome character. In fact relations at Edinburgh proved harmonious. Brewster was of course 78 in 1859, and had no doubt mellowed, but he was by no means a spent force, at least until his illness in 1864. It is tempting to say that as the first lay Principal Brewster was also, so to speak, the first Vice-Chancellor, the first to see the office in managerial and public-relations terms; but that role was reserved for his successor, Sir Alexander Grant. Brewster presided regularly over the university bodies, but did not really emerge as a public figure. His views on university reform appeared chiefly in his annual addresses to students, and in these we can find some clues to his opinions. All discussed the place of science in the university, and three distinct positions can be identified.

First, Brewster saw science as part of a general, liberal education—as he had already made clear at St. Andrews. Some knowledge of it was needed by all professional men and by those who played a part in public affairs. If politicians and legislators knew more about science, he claimed, they would not have neglected so scandalously the higher intellectual interests of the nation[8].

But science was also essential to a general education for less utilitarian reasons: along with the classics and philosophy, it was a means of discerning higher religious truths. Brewster believed in 'natural theology', in science as a confirmation of the evidences of Christianity and an illustration of the works of the Almighty. 'If these views be sound', he said in 1859, 'the instruction of literary and theological students, and indeed of the whole population, in the grand truths of the material world, becomes the duty of a Christian Church and a Christian State'[9].

And thirdly, Brewster put great emphasis on the usefulness of science and on technology. 'The history of civilization is the history of the applications of science and the arts to the material wants of our species', he said in 1864. His addresses to students tended to become rhapsodies in praise of the glories of modern technology—railways, steamships, telegraphs, photography, etc. And he insisted that the universities had a part to play in promoting such discoveries. The example of Watt at Glasgow was cited, but Brewster did not need to look so far afield. When he came to Edinburgh in 1859, this Museum had recently been founded as the Industrial Museum of Scotland, and the government had united its directorship with a chair of technology at the University, for which the Museum was to be a teaching aid. 'You will be able', Brewster told the students, 'to study the machinery, the instruments, and the various processes of art by which this country has become the workshop of the world, and the envy of all other nations'[10]. Unfortunately the professor, George Wilson, died in 1859, and the government then let the chair lapse. This was a blow to Brewster, who continued to campaign for the teaching of technology and engineering. A chair of civil engineering was endowed by the Dundee millionaire Sir David Baxter in 1868. The main innovation of the Brewster period, however, was the B.Sc. degree, the first in Scotland, which Edinburgh introduced in 1864. This was significant because it allowed the training of scientific specialists to bypass the traditional arts degree. But further developments had to wait on public or private money.

Many of Brewster's ideas were advanced ones even in the 1860s. How far did they fit in with those of other people? Were they part of any organised movement of opinion? First, one may say that his enthusiasm for technology, though striking, and already apparent in the evidence which he gave to the 1826 Royal Commission, was perhaps not exceptional[11]. The Scottish universities—unlike those in some countries—never showed any reluctance on principle to teach utilitarian subjects, and Glasgow acquired an engineering chair in 1840. But Brewster was perhaps a pioneer in his vision of technological humanism, his belief that the application of science to material wants was an aspect of civilisation, and had a rightful place in universities for that reason.

Secondly, his emphasis on science as part of the 'liberal' education of gentlemen was a bold one, especially in the 1830s. The Scottish universities maintained their traditional uniform arts curriculum down to 1892; it included mathematics and natural philosophy, but there was much resistance to bringing the generalist tradition up to date by embracing the more experimental branches of natural science. Natural history had a place in the standard curriculum at Aberdeen, and chemistry secured a more precarious foothold at St. Andrews, making use of the chair founded in 1840. But no such concessions were made in the two larger universities, perhaps because their scientific professors had plenty of work teaching medical students. It was not till the late 1860s and 1870s that a serious debate on the place of science in university culture got under way, and Brewster's belief in its liberal function anticipated the arguments used then by men like Huxley and Playfair, though he would no doubt have been deeply out of sympathy with the materialistic philosophy on which the claims of science were often based in later years.

Finally, one may doubt whether Brewster really subscribed to the idea of the university as a centre of original research. He campaigned throughout his career for state support for science, but usually saw this in terms of an Academy or Institute for researchers. His model was France, which had institutions of this kind, but (in the nineteenth century) a very under-developed university system. By mid-century, however, it was the example of Germany which was becoming more influential. One university reform movement with which Brewster was associated—he moved a resolution at its foundation meeting in 1853—was the Association for the Improvement and Extension of the Scottish Universities, which campaigned for more state aid and did much to bring about the 1858 Act. The leading spirit in this, the Edinburgh jurist James Lorimer, admired the German universities for their serious and solid learning, but like other university reformers of the time such as John Stuart Blackie he was interested primarily in historical, philological, legal, and theological scholarship, not in science; of eighteen new subjects in which the Association demanded the foundation of chairs, only three were scientific (geology, agricultural chemistry, and technology). In 1864 Brewster served on the committee of a short-lived revival of Lorimer's movement, the Scottish Universities and Educational Association, and as Principal he was active in seeking both public and private 'endowment' for the University. Even so, he was not really prominent in discussions of the question[12].

The German university model, like the Scottish one, gave a central role to the professor, but it saw him as a specialised scholar, taking the advancement of science or learning as his first duty, and training pupils to follow him in the pursuit of truth. That was not Brewster's ideal, or the pattern which his own career had followed. 'Successful teaching is the great function which must be performed in our Universities', he said in 1860; therefore 'in the choice of Professors ... we must seek for men who are able and successful teachers, and are willing to devote the whole energy of their minds to the faithful discharge of their duties'. The qualities demanded by teaching and research, he argued, were different, and they were best pursued in different institutions. This concept of the teaching university was in a good Scottish tradition, but it was becoming old-fashioned by the 1860s[13].

Thus although it is possible to piece together the fragments which compose Brewster's published views on university reform, one is left with a curiously negative feeling, that somehow his ideas never really engaged with other people's, and that he never fitted into any effective campaign or movement. Universities and university life were perhaps never central to the intellectual enterprise to which Brewster committed his energies for so many years.

Notes and References

1. *Evidence, Oral and Documentary, taken and received by the Commissioners ... for Visiting the Universities of Scotland, III. University of St Andrews* (P.P. 1837, xxxvii) 61, 77, 138.
2. D. Brewster, *Memoirs of the Life, Writings, and Discoveries of Sir Isaac Newton* 2 vols. (Edinburgh, 1855), II, 110.
3. *Report of the St Andrews' University Commissioners* (P.P. 1846, xxiii) 93–96.
4. *Ibid.*, ii.
5. *Edinburgh Evening Courant* (14 September 1859).
6. *The Miscellaneous Writings and Speeches of Lord Macaulay* (London, 1891), 709.
7. D. Brewster, *Introductory Address on the Opening of Session 1859–60* (Edinburgh, 1859), 6–8.
8. D. Brewster, *Introductory Address at the Opening of Session 1864–5* (Edinburgh, 1864), 7.
9. Brewster, *op. cit.* (7), 14–15.
10. *Ibid.*, 17; Brewster, *op. cit.* (8), 8.
11. *Evidence, Oral and Documentary op. cit.* (1), ... *I. University of Edinburgh* (P.P. 1837, xxxv) 556–561.
12. R. D. Anderson, *Education and Opportunity in Victorian Scotland. Schools and Universities* (Oxford, 1983).
13. D. Brewster, *Introductory Address at the Opening of Session 1860–61* (Edinburgh, 1860), 11, 14. Cf. his views in *Quarterly Review*, 43 (1830), 328–329.

Brewster as a Scientific Journalist

W. H. BROCK

Echoing the filial Mrs. Gordon's remark that 'editorial labours... occupied [my father] for a large portion of his life'[1], Edgar W. Morse has noted that 'Brewster's income depended upon his literary rather than his scientific efforts'[2]. Indeed, in many ways Brewster's career looked forward to those men of science in the 1860s and 1870s (like William Crookes and Norman Lockyer) who earned their livings, and who supported their scientific research, entirely from scientific journalism and the communication of science. No doubt if Brewster had not made major contributions to optics, or if he had been a less belligerent and controversial figure, historians of science would rank him alongside his slightly senior contemporaries, William Nicholson and Alexander Tilloch and others, who are better known for their journalism than for their contributions to the scientific enterprise.

Brewster's journalism started in 1799 when, at the age of seventeen, he began to contribute to the monthly *Edinburgh Magazine*, a literary miscellany founded in 1785, whose editor he became in 1802. Following its ignominious collapse in the following year and absorption into Archibald Constable's thriving *Scot's Magazine*, Brewster edited the latter for Constable until 1806 or 1807[3]. From 1808, when the first of its eighteen volumes appeared, until its conclusion in 1830, Brewster was the Editor and rather inefficient driving force behind one of *Britannica*'s competitors, *The Edinburgh Encyclopaedia*, a work compiled by 150 contributors which Brewster claimed was different from all its rivals by its originality and selectivity[4]. It was this vast journalistic enterprise which brought him into contact not only with the ambitious publisher William Blackwood, but also with the mineralogist Robert Jameson, whom he offered 5 guineas per sheet for the encyclopaedia article on mineralogy. Thus, by 1807, Brewster was working for the two principal publishing houses of Edinburgh, Constable (*Scot's Magazine*) and Blackwood (*Edinburgh Encyclopaedia*); and he had begun an apparently cordial relationship with Jameson, the Regius Professor of natural history.

Much remains to be done to capture the full variety of Brewster's journalism—ranging, as it did, from contributions to the *Encyclopaedia Britannica*, Jeffrey's and Napier's *Edinburgh Review*, Gifford's and Lockhart's *Quarterly Review*, the *North British Review*, Blackwood's *Edinburgh Magazine*, *Fraser's Magazine*, Hugh Miller's bi-weekly *Witness*, James Hogg's *Weekly Instructor*, Strahan's *Good Words*, Lardner's *Monthly Chronicle*, and *The Scientific Review*; and extending to his probable ownership of *The Perthshire Courier*, his abortive plan to launch a literary journal with Lockhart and Carlyle in 1825, or his encouraging positive support for the impoverished Carlyle at the beginning of his literary career[5]. Only one or two aspects of 'that large portion of his life' can be considered here.

To judge from the famous 'Chaldee Ms' squib which enlivened the seventh issue of *Blackwood's Edinburgh Magazine* in October 1817—the broadside which attacked Constable for stealing Blackwood's two editors, Pringle and Cleghorn, for the rival *Scot's Magazine*—both Jameson and Brewster had promised Blackwood their moral and literary support. The relevant hieratic verse personified Brewster as:

> 'a wise man which had a light in his hand and a crown of pearls upon his head... [who] said, Behold I will brew a sharp poison for the man which is crafty and his two beasts [i.e. Constable, Pringle and Cleghorn]. Wait yet till I come'[6].

Indeed, Brewster did write one or two articles for *Blackwood's Magazine*; but when Brewster asked Blackwood to help him launch a new quarterly journal of science, to be called *The Edinburgh Philosophical Journal*, as a competitor to Thomas Thomson's Glasgow-based *Annals of Philosophy* and Alexander

Fig. 5. David Brewster by William Bewick, 1824. Chalk drawing. (*Scottish National Portrait Gallery*).

Tilloch's London-based *Philosophical Magazine*, Blackwood had to refuse. As Blackwood related to John Murray:

> 'Dr. Brewster entered very fully into the plan of his journal; and from everything he said and showed me, I think he will make it a most interesting work. I pressed him to say what would be the terms he would expect as editor, and the rate of payment for contributions. He said this might be done in two ways—either by beginning with a small allowance, to be increased according to the sale, or starting at once with such an allowance both to the editor and contributors as would be proper to give, supposing the work successful: for the editor for each number'[7].

Brewster obviously pressed for the highest rate. Blackwood refused, and without any compunction or feelings for brand loyalty (for Blackwood was still carrying the burdensome *Edinburgh Encyclopaedia*), Brewster consigned the new journal to Constable.

The first issue of *The Edinburgh Philosophical Journal* 'exhibiting a View of the Progress of Discovery in Natural Philosophy, Chemistry, Natural History, Practical Mechanics, Geography, Statistics, and the Fine and Useful Arts', 'conducted by Dr. Brewster and Professor Jameson', appeared in June 1819. Each editor received £226—a princely sum, out of which, however, they each had to pay contributors[8].

The journal was not a success. As Carlyle noted in November 1819, '[Brewster's] Journal appears to be in a sickly state. Few speak of it; and those few without respect'[9]. Nevertheless its publisher, Constable, said nothing until March 1821, when he learned that Brewster and Jameson were at loggerheads. Two letters from Brewster to the chemist John Murray, suggest what was happening. In the first, written in November 1820, Brewster advised Murray as a prospective contributor to the journal:

> 'All communications for our Journal which are not on Mineralogy, Zoology or Botany [Jameson's provinces], should be sent to me, and will be carefully attended to.'

In a second letter, two months later, Brewster informed Murray:

> 'You mention that you sent a paper on Sulphuret of carbon. It was certainly never sent to me otherwise it would have been printed. I will thank you to recollect to whom you sent it that Enquiry may be immediately made for it, as the Paper is in my Department and *nothing* can be so injurious to the interests of a Journal, as to have it supposed that any communications from contributors are likely to be lost'[10].

Rightly or wrongly, Brewster appears to have accused Jameson of 'stealing' and losing papers that were his province to deal with in order to bias the character of the journal towards natural history.

By March 1821 Brewster and Jameson were not on speaking terms, and neither man was willing to make up the next issue of the journal. Constable, through his Manager, Mr. Cole, pointed out to both parties that the firm had invested a good deal of money in the journal and its advertisement, and how it had failed to render any profit after seven issues[11]. He went on to suggest that Napier, the editor of the *Edinburgh Review*, should arbitrate, threatening legal action against both editors if they refused to co-operate. Jameson suggested that a new periodical for natural history should be launched, implying, perhaps, that there was something in Brewster's suspicions; but Constable refused to alter the journal's general character and threatened Jameson with an injunction if he acted unilaterally, or abandoned the *Edinburgh Philosophical Journal* before the contracted twenty issues (5 years) had been completed[12]. Despite such threats, Constable did not succeed in getting the rival editors to meet Napier until the end of April[13].

Meanwhile, the grounds for Brewster's complaints against Jameson came out into the open. A more petty set of differences would be hard to find outside a school staffroom. Two of Brewster's eleven grievances will suffice to illustrate Brewster's jealousy of Jameson.

> '1st. That Mr. Jameson sent to the Printing Office a *Title*, and *Introductory Letter* to Scott's Narrative article not only in Dr. Brewster's department, but purchased by him; the Title bearing that this paper was *communicated by Dr. Traill* to Mr. Jameson, which was not the fact; and the Introductory letter purporting to be a letter from Dr. Traill to Mr. Jameson accompanying the communiciation, whereas Dr. Traill never wrote such a letter, but merely sent an Introductory Paragraph which Mr. Jameson converted into a letter. The obvious tendency of the step was to make Messrs. Constable & Co. and the public believe that this popular article, along with the two original and valuable ones by Major Rennell, were acquired by Mr. Jameson's influences, whereas they were acquired solely by Dr. Brewster's money...
>
> 2nd. That Mr. Jameson either sent back to the author or has detained for months a chemical paper without submitting it to Dr. Brewster; that Mr. Jameson has lost papers in Dr. Brewster's department'[14].

From this it seems clear that the demarcation of responsibilities between the editors was unworkable, and that each wanted chief responsibility for running the journal.

Jameson inconveniently lost Constable's letter containing Brewster's complaints, but his promise to make up his part of the next issue and, if necessary, to prepare the whole issue for press, no doubt began to put Jameson in a fairer light than Brewster in Constable's eyes[15]. In fact Napier must have succeeded in resolving this particular dispute, for the uneasy partnership continued until trouble flared up again in October 1822. At this point Constable, having based his calculations of profits against editorial expenses on sales of 2500 copies (that is, a run similar to that enjoyed by both the *Edinburgh* and *Quarterly Reivew*) began to express his serious worry about sales. By issue No. 8 (1821) he had reduced the print run to 2250, and to 2000 with No. 13 (September 1822). Since real sales, however, were between 1400 and 1700 copies per issue, printing and editorial expenses were out of step with sales returns by between £25 and £60 per issue, or £100–£200 per annum[16].

Hence when, after nearly five years of publication, it came the time to re-negotiate the terms of contract with Brewster and Jameson if the journal was to continue, Constable inevitably insisted that their remunerations would have to be calculated on sales of only 1250 copies. He also pointed out that he owned the copyright[17]. Reading between the lines at this point, it must have seemed obvious to both Brewster and Jameson that if the joint editorial fee was likely to be reduced by half it would be advantageous if one of them was appointed as sole editor. Was this why Brewster became so belligerent, accusing Constable quite unfairly of failing to support the journal, of failing to pay all his editorial expenses in the past, and of asserting his legal ownership of the journal which Brewster himself now claimed?[18].

By February 1824 the matter had become a serious legal dispute[19]. Obviously sick and tired of Brewster's tiresomeness, on 25 February Constable offered him some stark alternatives: Brewster could buy the backstock and copyright of the journal for £2100 (a sum Brewster could not have afforded at this time), or Brewster and Jameson could continue as editors under the new terms, or both could be dismissed with Constable continuing the journal under a new editor[20]. Furious, Brewster then unwisely attempted to set up an announcement in the April issue of the journal that he would be continuing the title with fresh publishers—an action which Constable immediately prevented by injunction[21]. Meanwhile Jameson, whatever may be said about his previous irrascible behaviour, agreed to Constable's reduced terms and found himself sole editor from 25 March 1824[22]. Under Jameson's editorship, and from 1826 under the modified title of *The Edinburgh New Philosophical Journal*, Constable's periodical survived into the 1860s, when it folded into William Crookes's exciting *Quarterly Journal of Science*.

Meanwhile Brewster had returned to Blackwood with his problem, and the latter, always happy to spite Constable, and perhaps hoping thereby to get Brewster to complete the *Edinburgh Encyclopaedia*, agreed to launch a rival imprint edited by Brewster entitled *The Edinburgh Journal of Science*. Despite Constable's copyright in the main title, this used the exact sub-title of the former joint journal[23], and appeared simultaneously on the bookstalls with the delayed April number of Jameson's journal. As Mrs. Oliphant sagely noted in the history of the firm:

'It was not like Mr. Blackwood's usual sagacity to believe that a man who had so neglected one publication [the *Encyclopaedia*] would be more diligent when he had two in hand. But the hopes with the new journal were equally fallacious. [Blackwood] was assured of a sale of 1250 and on this consideration agreed to pay the editor £100 or £115 for each number, besides paying ten guineas a sheet to the contributors'[24].

In other words, Blackwood accepted the very terms he had sensibly declined in 1819: of course, Blackwood soon discovered that he could scarcely sell 1000 copies, and that there were to be no profits. Mrs. Oliphant continues:

'Dr. Brewster when appealed to, would neither release the publisher nor exert himself more diligently. He left Edinburgh calmly, like all the other people connected with the University, for six months of the year, and drove the printing house frantic with incessant delays.'

In 1826 Blackwood was forced to take legal action against Brewster to force him to produce copy on time—no doubt the source of Carlyle's interesting remark that year that:

'The Doctor is in the blackest humour about "the badness of the times"; as in truth he has some reason to be, being involved in lawsuits with his booksellers, perplexed with delays in his Encyclopaedia, and finding publishers so shy of embarking in any of his schemes'[25].

Elsewhere, Carlyle also makes it clear that Brewster lived well beyond the means of a non-academic who supported himself solely through literary earnings[26]. Similar in format and content to its rivals, the *Annals of Philosophy*, *Philosophical Magazine*, and the *Edinburgh New Philosophical Journal*, though without their variety, Brewster's journal survived until 1832, becoming chiefly known and read for its strong views on the alleged decline of science in Great Britain.

In 1827 Blackwood sold his interest in Brewster's journal to his London agent, Thomas Cadell, who, with Brewster's aid, managed to find additional proprietorial help in Edinburgh and Dublin[27]. Hence, by April 1829 when this triumvirate system was instituted, Brewster's *Edinburgh Journal of Science* was, in effect, an Edinburgh, London and Dublin *Journal of Science*, lacking only editors in all three capitals. Possibly it was this arrangement which, following the successful formation of the British Association for the Advancement of Science at York in September 1831, inspired Brewster to think of a national commercial science journal[28]. As he confided to the Rev. William Vernon Harcourt:

'I wish much to have your opinion of *two objects* which I think can be accomplished by the British Association and which would tend much to promote British science.
   1. The first is to have a *monthly* journal of science called the *British Journal of Science* published in

London in place of the *four journals* which are now in existence. All that the Association would do, would be to patronise it. All the *four* scientific journals are I believe carried on without any remuneration, and the editors who [?] have therefore no motive for exertion. In order to save themselves from loss the editors are obliged to insert popular papers, likely unsuited to such works; and in consequence of this the editors being acquainted with only one branch of science, their numbers abound with the most contemptible articles that disgrace British science in foreign countries. In order to make a good scientific journal, all the best papers from the four different journals would be required.

Though I receive no remuneration whatever [*sic*] from my *Journal*, I do not intend to discontinue it, but I would willingly do so in order to carry into effect the plan I mention; and I should have no objections if required to take a department in the general journal. All the *four* editors might do the same, and thus we should have one good journal made out of the whole. If you view this in the light I do, I would write to Mr. R. Taylor, the editor of the *Annals of Philosophy*[28] and ascertain his sentiment, and if they are favourable, application might be made to the other editors. All this might be done independent of the Association but, with their approbation and aid, it would be done more efficaciously'[29].

Although nothing came of Brewster's impossibly grand scheme, contact with the printer and publisher Richard Taylor, the sole owner of the reputable and internationally read *Philosophical Magazine*, led to the sale and amalgamation of Brewster's journal in 1832. Jameson crowed triumphantly and prematurely to his nephew:

'Brewster's Journal has died a natural death, so that Sir David must now take up some other scheme. It must be mortifying to him after all the abuse he heaped upon the Editor of the New Philosophical J. [i.e. Jameson himself] to be forced from a field which he considered as his own'[30].

Under the new system Brewster became one of the three (later four) editors of the *London and Edinburgh Philosophical Magazine*[31], a position he filled honourably and conscientiously until his death in 1868. As proprietor, managing editor and printer, Taylor seems to have ensured that there was no frictional jealousy such as had occurred between Brewster and Jameson; nevertheless, in 1852 Brewster could still claim that 'we have not in England, or Scotland, or Ireland a single Journal of science, or Magazine of Art of the least merit—either well conducted, well illustrated, or well circulated'[32].

As this episode suggests, Mrs. Gordon's quotation of Fraser and Blackie, two successive editors of the *North British Review* who praised Brewster for the punctual way he delivered copy and for his conscientiousness, should be taken with a pinch of salt[33]. E. J. Shattock has shown in her study of the *North British Review* that they thought very differently during Brewster's lifetime[34]. No doubt by 1869 their memories had dimmed and they wished to be charitable to the devoted daughter and follow the conventions of Victorian biography. However, in practice Brewster traded very much on the fact that he and Thomas Chalmers had been instrumental in launching the Free Church's *North British Review* in 1844. Although Brewster's contributions ensured that the *Review* contained a slightly higher proportion of science articles than either the *Edinburgh* or *Quarterly Review* in the period 1844 to 1860, his relationship with the editors was frequently difficult.

'Brewster regarded his relationship with the review as privileged, and relentlessly pressed upon Fraser [Editor 1850–57] articles of over 50 pages on subjects of his *own* choice for every issue. He resisted all pleas for a reduction in length, insisting that no previous editor had ever complained, and added to the general irritation by complaining that he had been dealt with unfairly financially —particularly galling to Kennedy [the publisher] as he was paid at a higher rate [of £15 per sheet] than any of the other contributors'[35].

In concentrating on Brewster's career as an editor and journalist, this essay has inevitably tended to reveal the less attractive side of a man of science struggling to earn his living. However, this image of Brewster must be balanced by the fact that he was almost invariably a splendid writer, and it was the extraordinary power and readability of his prose which allowed editors to put up with his various delinquencies, and which makes him still such a pleasure to read today. Consider, for example, his splendid ridicule of the trustees of the Earl of Bridgewater:

'The dedication of nearly *ten thousand pounds* to the composition of a work "on the Power, Wisdom, and Goodness of God, as manifested in the Creation", is an event without any parallel in the history of our literature. A bequest of such munificence and piety could not fail to inspire us with gratitude to its liberal donor; and though it was calculated to excite the highest expectations of a work, for the execution of which such ample means were provided, yet these feelings were alloyed with extreme anxiety respecting its judicious appropriation. We regret to say, that our anxiety has been justified, and our expectations disappointed. We had counted upon worshipping the creative Spirit in one massive Temple, whose materials had been prepared by the wisdom and industry of past ages, and at whose altar there would minister some High-Priest, whom piety and science had combined to consecrate; but we have been summoned to an interrupted and multifarious oblation, and have been directed to eight separate shrines, in which as many votaries are offering up their insulated orisons'[36].

Ignoring Brewster's pique at not having been chosen and rewarded for writing a Bridgewater Treatise himself[37], Brewster's comment was well made and the lively bantering style encouraged the reader forward into the real business of the essay—a highly critical review of Whewell's *Astronomy and General Physics*.

Brewster's highly visual style depended very much for its power upon sustained metaphor, as in the following two, very different passages on Wheatstone's electromagnetic master clock for observatories, and on palaeontology.

'Nor is this invention confined to observatories and large establishments. The great horologe of St. Paul's might by a suitable network of wires, or even by the existing metallic pipes of the metropolis, be made to command and regulate all the other steeple-clocks in the city, and even every clock within the precincts of its metallic bounds. When railways and telegraphs extend from London to the remotest cities and villages, the sensation of time may be transmitted along with the elements of language; and the great cerebellum of the metropolis may thus constrain by its sympathies, and regulate by its power, the whole nervous system of the empire'[38].

'Geology, Zoology, and Botany should be most carefully and completely treated... They form, indeed, the key to the hieroglyphics of the ancient world; they enable us to reckon up its almost countless periods; to replace its upheaved and dislocated strata; to replant its forests; to reconstruct the products of its charnel-house; to repeople its jungles with their gigantic denizens; to restore the condors to its atmosphere, and give back to the ocean its mighty leviathans. And such is this force with which these revivals are presented to our judgement, that we almost see the mammoth, the megatherion, and the mastodon, stalking over the plains or pressing through the thickets; the giant ostrich leaving its foot-writing on the sands; the voracious ichthyosaurian swallowing the very meal which its fossil ribs enclose; the monstrous plesiosaurus paddling through the ocean, and guiding its lizard-trunk, and rearing its swan-neck, as if in derision of human wisdom; and the pterodactyle, that mysterious compound of birds, and brutes, and bats, asserting its triple claim to the occupancy of earth, ocean, and atmosphere'[39].

Similarly, who but Brewster could have compared railways to the Iliad, as he was to do in 1849?[40].

In sum, our portrait of Brewster as a science journalist has both good and bad points. He was undoubtedly worth reading himself, and useful to editors for the contacts he could provide with other eminent potential writers. Carlyle always found him a reasonable man[41], and it must always be remembered that Brewster's disputatious, belligerent, combative, self-centred mannerisms were common coin in literary Edinburgh during the first three decades of the nineteenth century. As an editor there is evidence that he took pains to advise authors how to improve and to revise their work, acting in fact as their referee[42]; and, as Shattock has strongly emphasised, 'he approached... the "holy alliance" of science and revelation with an aggressive conviction of which the Church could be proud' and which strongly appealed to editors and readers of the large Evangelical press[43].

Notes and References

1. Margaret M. Gordon, *The Home Life of Sir David Brewster* (Edinburgh, 1869), 99.
2. E. W. Morse, 'David Brewster' in C. C. Gillispie (ed.), *Dictionary of Scientific Biography* 16 vols. (New York, 1970–76), II, 451–454.
3. Gordon, *op. cit.* (1), 99 incorrectly states that the *Edinburgh Magazine*—defunct in 1803—became the *Edinburgh Philosophical Journal* in 1819. To avoid confusion, the following bibliographical note may be helpful. The *Edinburgh Magazine* was founded and edited by the Edinburgh bookseller James Sibbald in January 1785. In 1791 the ownership passed to Laurie & Symington and the editing to Dr. Robert Anderson. Brewster was editor from 1802 until the journal's merger with Constable's *Scots Magazine* in December 1803. *Blackwood's Edinburgh Monthly* (April 1817–September 1817) became *Blackwood's Edinburgh Magazine* in October 1817, the reference to Edinburgh being dropped only in 1906.
4. For Brewster's views on encyclopaedias, see his review of the 7th edition of the *Encyclopaedia Britannica* in *Quarterly Review*, 70 (1842), 44–72.
5. For Brewster's contributions to the *Edinburgh Review*, the *Quarterly Review*, *North British Review*, *Fraser's Magazine* and *Monthly Chronicle*, see W. E. Houghton (ed.), *The Wellesley Index to Victorian Periodicals, 1824–1900*, 3 vols. (Toronto, 1966). His other journalism is mentioned in Gordon, *op cit.* (1), or in the Brewster files held in Edinburgh University Library and the National Library of Scotland.
6. F. D. Tredrey, *The House of Blackwood* (Edinburgh, 1954), 249.
7. Mrs. Margaret Oliphant, *Annals of a Publishing House: William Blackwood and his Sons. Their Magazine and Friends*, 3 vols. (Edinburgh and London, 1897–98) II, 8.
8. National Library of Scotland, Ms 790 f. 513.
9. C. R. Sanders and K. J. Fielding (eds.), *The Collected Letters of Thomas and Jane Welsh Carlyle* (Durham, North Carolina, 1970–) I, 208 (18 November 1819).
10. Edinburgh University Library, AAF Brewster 20, Brewster to Murray, November 1820 and 26 January 1821.
11. National Library of Scotland, Constable Papers 491 f. 260 (12 March 1821).
12. *Ibid.*, 491 f. 262 (13 March 1821); 491 f. 266 (15 March 1821).
13. *Ibid.*, 491 ff. 275, 277, 305 (19, 30 March, 27 April 1821).
14. *Ibid.*, 491 f. 280 (4 April 1821).
15. *Ibid.*, 491 f. 294 (17 April 1821); 491 f. 300 (26 April 1821).
16. *Ibid.*, 791 f. 629 (4 October 1822); 791 f. 642 (10 October 1822).
17. *Ibid.*, 492 f. 189 (16 January 1824).
18. *Ibid.*, 492 f. 201 (5 February 1824).
19. *Ibid.*, 492 f. 209 (12 February 1824).
20. *Ibid.*, 492 ff. 221, 228 (25 February and 4 March 1824).
21. *Ibid.*, 492 f. 232 (10 March 1824).
22. *Ibid.*, 492 f. 245. Jameson was to be paid £100 per issue, out of which he had to pay for postage and all translations, *ibid.*, 492 f. 252.

23. See Brewster's polemical 'Preface' to the first issue of *The Edinburgh Journal of Science*, *1* (1824).
24. Oliphant, *op. cit.* (7), II, 8.
25. Sanders and Fielding, *op. cit.* (9), IV, 166 (9 December 1826).
26. *Ibid.*, IV, 177 (1 January 1827).
27. T. Bestermann (ed.), *The Publishing Firm of Cadell & Davies. Select Correspondence and Accounts 1793–1836* (Oxford, 1938), 82–83.
28. I.e. *Philosophical Magazine or Annals of Chemistry, Mathematics, Astronomy, Natural History and General Science* which Richard Taylor formed from the amalgamation of Thomson's *Annals of Philosophy* with his *Philosophical Magazine* in January 1827.
29. Brewster to Harcourt, 1 February 1832, *The Harcourt Papers* (Oxford, 1880–1905). I thank Mr. J. B. Morrell for lending me a copy of this letter. Although there were more than four journals in existence, Brewster presumably had in mind the four monthlies *Edinburgh New Philosophical Journal*, *Edinburgh Journal of Science*, *Philosophical Magazine* and probably J. C. Loudon's *Magazine of Natural History*. Brewster's second proposal involved the amalgamation of the *Transactions* of the Royal Societies of London and Edinburgh. See Jack Morrell and Arnold Thackray, *Gentlemen of Science* (Oxford, 1981), 64–65.
30. Robert Jameson to Thomas Jameson, 3 August 1832: Edinburgh University Library, Ms Gen. 1996/2/17.
31. From 1840 *The London, Edinburgh and Dublin Philosophical Magazine*.
32. [D. Brewster], 'Prince Albert's Industrial College of Arts and Manufacturers', *North British Review*, *17* (1852), 519–558, 554.
33. Gordon, *op. cit.* (1), 179–180.
34. E. Joanne Shattock, 'A Study of the *North British Review* (1844–1871); Its History, Policies and Contributors', unpublished Ph.D. thesis (University of London, 1973). I am grateful to my colleague, Dr. Shattock, for allowing me to consult, and quote from, this thesis.
35. *Ibid.*, 148. For Brewster's contributions see note 5 above.
36. [D. Brewster], 'The Bridgewater Bequest', *Edinburgh Review*, *58* (1833–34), 423–424.
37. W. H. Brock, 'The Selection of the Authors of the Bridgewater Treatises' *Notes and Records of the Royal Society*, *21* (1966), 162–179, esp. 168–169.
38. Brewster, *op. cit.* (4), 57.
39. *Ibid.*, 58.
40. [D. Brewster], 'The Railway System of Great Britain' *North British Review*, *11* (1849), 570.
41. Carlyle, *op. cit.* (9), V, 65 (27 January, 1830).
42. *E.g.* Brewster to Murray 30 December 1825; Edinburgh University Library, AAF Brewster 23.
43. Shattock, *op. cit.* (34), 131.

Brewster, Evangelism and the Disruption of the Church of Scotland

PAUL BAXTER

David Brewster was a fervent advocate of science and technology as instruments of human progress and of scientific education as an enlightening and civilising influence in society. He was also a staunch champion of the Evangelical wing of the Scottish Church. Between the two causes he perceived no conflict, and indeed believed that the forces of orthodox Calvinism and of science could work together for their mutual benefit. In this paper I shall discuss the ways in which Brewster forged links between his scientific beliefs and his theological and moral commitments. I shall suggest that to a considerable extent Brewster's attitudes were shared by other Scottish Evangelical scientists with whom he was in close contact. However I shall also highlight some aspects of his approach to the problem which seem to have been peculiarly his own, and in my conclusion I shall speculate as to the reasons for this individualistic streak in Brewster's work.

The word 'Evangelical' was a term with a precise meaning in Scotland during the early nineteenth century. It identified one of the two major parties within the Established Church, the other being the so-called Moderate party. Their differences were partly doctrinal, partly concerned with matters of Church government; indeed, since Scripture was sometimes invoked in debates about ecclesiastical polity, it is difficult entirely to separate the two areas. A divided Church was not peculiar to Scotland, for undoubtedly differences of outlook also existed in the Church of England at that time, but the conclusion to many years of party strife had no exact English counterpart. In 1843 a schism occurred in the Church, known as the Disruption, leading to the formation of the Free Church of Scotland, which, uniting with a number of earlier breakaway groups, remained outside the national Church until 1929[1].

The early decades of the nineteenth century saw the revival of the Evangelical party's fortunes and the corresponding decay of the Moderatism which had been the dominant force during the flowering of Scottish intellectual culture in the late eighteenth century. Evangelical divines attacked what they considered to be the laxity of the Moderates over matters of doctrine and discipline, and in their preaching restored the Calvinistic emphasis on man's depravity and the regenerating power of the Gospel in place of the moral discourses favoured by their Moderate brethren. The immediate issue which led to the Disruption was the question of popular rights in the choosing of ministers. More specifically it concerned the right of a congregation or its representatives to reject the ministerial candidate chosen by the patron, the person or institution with the right to nominate the holder of the benefice. The Evangelicals held that there was a right sanctioned by law and tradition to resist the intrusion of an unpopular presentee, and hence became known as the 'non-intrusionists'. Some went further than this, transforming the demand for a congregational right of veto into a more radical call for the complete abolition of patronage. The logical consistency of their position was perhaps borne out after the passing of the Veto Act of 1834, which, whilst guaranteeing on paper the rights of congregations, led only to head-on collisions between the Church Courts and the Civil Courts, the Civil Courts invariably siding with the aggrieved presentee. After nearly ten years of strife, about 450 of the Church's 1200 or so clergymen decided that they could remain in the State Church no longer. The parting of the ways took place at the General Assembly of 1843, when the Free Church of Scotland was born.

The Evangelicals' first commitment was to spreading the Gospel message, both at home amongst the 'infidel masses' living in poverty and wretchedness in the poorest districts of Edinburgh and Glasgow, and overseas amongst the heathen of Africa and Asia.

Fig. 6. The Signing of the Deed of Demission by David Octavius Hill. 1843–1867. Oilpainting. *(Free Church of Scotland)*.

Yet they were by no means backwoodsmen of science and secular learning, numbering in their ranks several laymen and even a few divines distinguished in other than theological pursuits. In no sense were they on the margins of Scottish intellectual and cultural life.

Although according to the biography by his daughter Maria, his deepest religious convictions took root late in his life—during the 1850s in fact—Brewster identified from very early in his life with the Evangelical cause[2]. In 1806 he wrote a pseudonymous pamphlet taking the side of Professor John Leslie when the Moderate party sought to prevent Leslie's election to the mathematics chair at Edinburgh University[3]. Originally destined for the ministry himself, Brewster gave up this ambition apparently because of his extreme nervousness of public speaking. During the so-called Ten Years Conflict between the passing of the Veto Act and the Disruption, Brewster wrote on a number of occasions to Lord Brougham advocating the cause of the non-intrusionists and urging Brougham to do something about the constitutional position of the Church. He also took part in the procession of seceding churchmen in 1843 and was afterwards an elder in the Free Church. Indeed his allegiance to the new Church almost made him a victim of ecclesiastical intolerance when the Established Church attempted unsuccessfully to unseat him from his position as Principal of the United College of St. Leonard and St. Salvator at St. Andrews. The ground on which they sought to do this was the requirement in the Test Act that university professors must sign the Westminster Confession of Faith pledging allegiance to the Church of Scotland and its doctrines. Although this was widely regarded as an anachronism and had fallen into disuse, it was revived after the Disruption as a device for removing Free Churchmen from their posts.

In this paper I shall be concerned mainly with Brewster's popular scientific writings, covering a wide range of subjects. I draw heavily on his articles for the *Edinburgh Review* and later for the newly-formed *North British Review*, a periodical broadly sympathetic to the Free Church. I shall not deal with his optical researches, although it has been argued in a thesis by E. W. Morse that Brewster's opposition to the undulatory theory of light sprang from his conviction that speculations concerning the existence of a luminiferous ether were unwarranted and could encourage scepticism[4].

Brewster was not the only standard-bearer for science in the ranks of the Free Church. Contemporary writers often mentioned three other names when emphasising its achievements in secular learning and defending it against accusations of being hostile to the diffusion of knowledge. These three were the Rev. Thomas Chalmers, Hugh Miller and the Rev. John Fleming. Chalmers, remembered for his evangelical work amongst the lower classes in St. John's parish in Glasgow, was in many ways the architect of the Free Church and its leader until his death in 1847. He was also greatly interested in science, having once intended to embark on a scientific career himself, and enthralled Glasgow congregations with his astronomical discourses, which sought to demonstrate that Christianity had nothing to fear from recent astronomical discoveries[5]. Hugh Miller was something of a folk-hero, the self-made man from Cromarty, a fishing village in north-east Scotland, who had progressed from stonemason to geologist, making important discoveries in the Old Red Sandstone formation. He was also a writer of powerful polemics on ecclesiastical questions and in 1840 he became editor of the Evangelical newspaper the *Witness*[6]. John Fleming combined a career in the ministry with distinction in geology and zoology. In 1845 he became professor of natural science at the Free Church's New College in Edinburgh[7]. Brewster, Chalmers, Miller and Fleming all wrote extensively about the relationship between science and Christian belief. In this section I want to discuss some common strands running through their work, and to try to explain their occurrence.

In a society in which science was not an autonomous profession, such as the society of nineteenth century Britain, those engaged in scientific pursuits were under considerable pressure to show that their chosen field of endeavour could help to uphold the theological and moral order. There were a number of different strategies for making such connections and some proved more successful and durable than others.

Geology, as a relatively new science and one whose aims were from the outset treated with some suspicion in religious circles, was particularly in need of such a link. For geologists one possible linking strategy was to attempt to harmonise the geological record with the account of the formation of the earth given in the book of Genesis. Very few scientists of any repute undertook this project in its most extreme form which, according to the strict Biblical chronology, compressed all geological processes into the span of about 6,000 years[8].

A strategy which enjoyed greater popularity for a time was the collection of geological evidence for the occurrence of the Great Flood described in Genesis. The so-called diluvial theory, advocated by, amongst others, the Oxford geologist William Buckland, attributed the gravel deposits widely distributed over the northern hemisphere and the form of many hills and valleys to the action of a single, recent and universal flood[9].

Brewster, like other Evangelical writers, despised such attempts to mingle science, and particularly the geological record, with the Scriptures. As early as 1814 Chalmers had adopted the so-called interval theory which gave geologists a high degree of freedom to work unfettered by the Biblical chronology[10]. The theory suggested that Scripture was silent about events which had occurred between the first and second verses of the first book of Genesis and thus admitted that the Bible might describe only the most recent of a whole series of creations, spread over a vast period of time. It is also interesting to note that one of Buckland's keenest adversaries over the diluvial theory was John Fleming, who maintained that the Biblical Flood had been an entirely tranquil affair bringing about no major geological changes[11]. For Brewster, the Evangelicals' defence of the independence of geology and Scripture was a matter for considerable pride, and an opportunity to remind his readers of previous martyrs in the cause of scientific advance:

> '... the highest demands of truth, and the best interests of mankind, are invariably sacrificed when religion is intruded into questions of science and civil policy. Prejudice is then arrayed against knowledge; and reason stands the shackled victim of ignorance and fanaticism. The persecution of Galileo for maintaining doctrines which had been previously demonstrated by a pious and exemplary ecclesiastic, is fraught with deep instruction to every friend of religion; but a still more impressive lesson is now read to them in the recent triumphs of geological discovery'[12].

In the early nineteenth century the most important link between the realm of science and the realm of theology was provided not by schemes of Biblical reconciliation but by the enterprise of natural theology. Natural theology took religious worship outwards from the recesses of the human spirit and the confines of church and chapel to the boundless realm of external nature. It sought to use the face of the external world to establish the existence of a Creator, and to provide proofs of His benevolence, wisdom and power. Any branch of science could, in principle at least, become the handmaiden of natural theology, although in practice the main emphasis tended to be on the life sciences: on the forms of animal bodies and of plants, and on the precise adaptation of each organism to its environment. Nevertheless the authors of the eight treatises produced under the terms of the Earl of Bridgewater's bequest and published during the fourth decade of the nineteenth century, wrote on topics as diverse as astronomy and general physics, and the social and economic system. As Professor Gillispie and others have emphasised, natural theology strongly pervaded the work of many leading British scientists of the first half of the nineteenth century. Its promise was that scientific researches would throw light on the character of the Creator, and thus make science the friend of orthodox Christian belief rather than its foe. It could also serve to vindicate the ways of God to man, explaining away apparent defects in the social order as God-given and therefore immutable[13].

The Scottish Evangelicals reposed far more confidence in natural theology than they did in schemes of Biblical reconciliation, and indeed Chalmers accepted the invitation to write the *Bridgewater Treatise* dealing with the social and economic system[14]. For Brewster the principles of natural theology served to justify repeated calls for better and more widely available education, especially scientific education. The life sciences, in particular, had a valuable religious purpose:

> 'While the vulgar gaze in mysterious wonder at the results of creative power, the student of nature perceives the unity of design and of purpose which pervades the whole; and he is permitted to trace the steps and pursue the laws by which the Omniscient Spirit has accomplished his Work'[15].

Natural theology did have a number of potential disadvantages. It had nothing to say about crucial points of Christian doctrine: about the Fall of Man and the process of redemption, to which the Evangelicals were strongly committed. Some natural theologians also left unresolved the problem of how to respond to developments in science. In particular, was the explanation of phenomena in terms of natural laws of increasing generality a tendency to be welcomed or worried about? Did it elevate our conception of the Creator to discover that general laws explained what we had previously thought to be individual instances of the divine wisdom? Or did it push the Creator so far back from His Creation that it became easier to argue that the universe was ruled by chance or by necessity? When Evangelical writers discussed the subject in general terms they invariably came down on the side of a busy, resourceful Creator rather than a remote ruler whose plans were unfolded by means of unvarying laws. Nevertheless when it came to evaluating particular scientific theories, the Evangelicals were much less guarded, at least until a watershed event in 1844 which I shall discuss later. A striking example of this fairly relaxed attitude is provided by their reception of the nebular hypothesis in astronomy[16].

The nebular hypothesis had been used by Laplace as an explanation of the origin of our own solar system and by Sir William Herschel to help him make sense of his observations of the nebulae, or luminous patches, in the heavens. The hypothesis held that stars and their attendant planets had been formed by a process of condensation from a rotating mass of incandescent gas. Laplace believed that the theory could account qualitatively for many of the observed regularities in the solar system such as the motion of the planets in the same direction and in nearly the same plane around the sun[17]. Herschel suggested that those nebulae which successively more powerful telescopes had failed to resolve into clusters of stars might not consist of stars at all but might instead be such incandescent clouds. He also hypothesised that the so-called nebulous stars in each of which a star appeared embedded in a nebulous cloud, might be intermediate between true nebulae and fully-formed stars[18].

The theory gained considerable popularity in Scotland during the late 1830s and early 1840s, helped by the works of John Pringle Nichol, professor of practical astronomy at Glasgow University[19]. Evangelical writers praised Nichol's books and lectures, and the theory had no more enthusiastic supporter than Brewster himself. In 1838 Brewster dwelt at some length on the hypothesis in a review of Auguste Comte's *Cours de Philosophie Positive* for the *Edinburgh Review*. Brewster assured his readers that the theory represented no threat to the need for an active Creator, and even made a further concession to the widening domain of natural law, which he must later have regretted:

> 'But even if science could go infinitely farther, and trace all the forms of being to their germ in a single atom, and all the varieties of nature to its developement [sic], the human mind would still turn to its resting-point, and worship with deeper admiration before this miracle of consolidated power'[20].

In 1844 an anonymously-published book called *Vestiges of the Natural History of Creation* attempted to extend the domain of natural law much further. It was the work of the Edinburgh bookseller and publisher Robert Chambers and it argued that the whole of nature, organic and inorganic, was under the rule of a law of development. The book was in a sense a forerunner of Darwin's *Origin of Species*, although the evolutionary mechanism postulated was closer to that of the French naturalist Jean Baptiste de Lamarck. Chambers drew evidence for his theory from a wide variety of scientific subjects in which he had read voraciously if not deeply. From astronomy he seized on the nebular hypothesis, holding the development of a star system from a gaseous cloud to be a kind of inorganic parallel to the development of complex life forms out of simpler ones. Chambers also argued that this image of creative processes gave a nobler view of the Creator than the traditional one which had required individual acts of creation for each new species. In other words he claimed that his work was a perfectly legitimate addition to natural theology[21].

Vestiges brought down practically the united wrath of the scientific elite, the Scottish Evangelicals being amongst its sternest critics. Brewster and Miller joined other opponents like the Cambridge geologist Adam Sedgwick in condemning the book for fostering irreligion and in ridiculing its factual errors. In a long and vitriolic attack in the *North British Review* Brewster described the author as 'a naturalist from books and not from observation'. The work was 'prophetic of infidel times', and indicated 'the unsoundness of our general education'[22].

Unlike other opponents, the Evangelicals made the additional move of immediately abandoning the nebular hypothesis. This about-face was most dramatic of all in the writings of Brewster himself, who in his review of *Vestiges* devoted several pages to a refutation of the nebular theory which he had been ardently advocating in the same journal nine months previously. Although recent discoveries by the astronomer Lord Rosse had weakened the foundations of the theory, it can be shown that these discoveries in themselves did not cause Brewster's change of heart because he had written an article favourable to the theory *after* they had been announced[23]. Indeed apart from Rosse's discoveries all the arguments used against the theory by Brewster could equally well have been used before 1844.

In the light of *Vestiges* Brewster described the nebular theory as a hypothesis 'so improbable in its very nature, and so gratuitous in all its assumptions', taking advantage of the reviewer's anonymity and saying nothing about his past attitude to the theory[24]. He did subsequently admit that he had once supported the theory and stressed that *Vestiges* had caused him to change his mind[25]. Thereafter he made regular assaults on the nebular hypothesis with all the zeal of a convert repenting of previous heresy. Indeed after 1844 the Evangelicals were much more suspicious of all such attempts to substitute natural law for individual acts of the Creator. Although he was advanced in age by the time of its publication it is not surprising to find that Brewster wrote a hostile review of the *Origin of Species* for a religious periodical[26].

Brewster resembled other Evangelical writers in his distrust of all schemes for reconciling the Bible and science, in his early enthusiasm for the nebular hypothesis, and in his later wholehearted rejection of such cosmological speculations. His initial support for the nebular hypothesis had probably been more fervent than that of other Evangelicals but there were other respects in which his attempts to build bridges between science and religious belief were more distinctive. One such area was Brewster's enormous enthusiasm for a belief in the so-called plurality of worlds, the notion that other worlds were inhabited. This was not a new idea, but its significance was increased by the discoveries of Sir William Herschel, in the late eighteenth century. Herschel's telescopes revealed vast numbers of new stars, and the notion that each one of them might be surrounded by planets, the abodes of intelligent beings, was a staggering prospect to some. It was to combat the suggestion that a plurality of worlds was in conflict with the Christian scheme of redemption, that Thomas Chalmers preached his series of *Astronomical Discourses* which I mentioned earlier.

Chalmers' concern was entirely with overcoming objections to orthodox belief, rather than with a defence of the plurality of worlds itself. Brewster, on the other hand, seems to have been warmly enamoured of the idea that other planets in the universe might be the homes of intelligent beings. In 1853, he was incensed by the publication of an anonymous essay which argued against the doctrine. The essay was by the Cambridge scientist, William Whewell, with whom Brewster had previously crossed swords, but we know that even before Brewster knew who had written it, he was appalled by its contents[27].

Brewster wrote a hostile review for the *North British* and later expanded his attack into a full length book, called *More Worlds than One: The Creed of the Philosopher and the Hope of the Christian*[28]. As its title suggests, the book is a strange one. It mingled a scientific argument about the plausibility that life existed on other worlds with the suggestion that the planets were a kind of physicalised heaven. According to Brewster the Christian looked to the stellar systems 'as the hallowed spots on which his immortal existence is to run'[29]. In our own solar system he believed that all the major planets were inhabited, and he suggested that the satellites, including our own moon, and even the sun might also be furnished with living occupants. In relation to the sun he revived the suggestion of Sir William Herschel that its heat derived from an outer region of self-luminous clouds and that the inner region was cool. By analogy with our own system every star might have its family of inhabited planets. Indeed since God made nothing in vain, Brewster felt that they *must* be inhabited. Moreover in a crude equation of size with moral and mental ascendancy he believed that the sun—'a domain so extensive, so blessed with perpetual light'—was occupied by 'the highest orders of intelligence'[30].

I suggest that the dual character of these other worlds in Brewster's cosmology indicates two quite different roles that they could play for an orthodox Calvinist. On the one hand they served to give astronomy a new theological and moral significance, filling something of a gap that was left by the discarding of the nebular hypothesis. Through his telescope the astronomer now saw not vast creative processes in operation but, nobler and loftier still, glimpses of heaven itself. At the same time, by admitting the possibility that the inhabitants of other worlds might be in a position morally far advanced on our own, the belief enabled a scientist deeply aware of human corruption to speculate about the blessings that could flow from science and technology, given a change in the moral climate in which they were conducted. Brewster asked:

'What inconceivable and countless functions may we not assign to that plurality of intellectual communities, which have been settled, or are about to settle, in the celestial spheres? What deeds of heroism, moral, and perchance physical! What enterprises of philanthrophy,—what achievements of genius must be required in empires so extensive, and in worlds so grand!'

'... may there not be a type of reason of which the intellect of Newton is the lowest degree? May there not be a telescope more penetrating, and a microscope more powerful than ours?—processes of induction more subtle,—of analysis more searching,—and of combination more profound? May not the problem of three bodies be solved there,—the enigma of the luminiferous ether unriddled,—and the transcendentalisms of mind embalmed in the definitions and axioms and theorems of geometry?'[31].

Brewster also engaged in some rather unconventional geological speculations in support of his favourite theory. Whewell's argument against inhabited worlds had depended heavily on analogies between time and space. He used the vastness of the geological timescale and the very recent appearance of man on earth to suggest that the localisation of intelligent life in space became more plausible once it was realized how *short* our occupancy of the earth had been. Brewster did all he could to overturn this analogy by trying to shorten the geological timescale. To suggest that geological causes had operated with greater intensity in the past was not at all at odds with what many geologists believed, but Brewster went further than this, suggesting that the Creator might have suspended the action of secondary causes altogether and intervened directly in the physical processes which had brought the earth to its present state:

'Under the influence of electric agency, and chemical and physical forces of higher activity, even secondary causes may have operated much more quickly than at present; but as creative power must have, at some period, acted by its mighty fiat, and actually did, even in the opinion of geologists themselves, by the *direct creation* of new life, after all pre-existing life had been destroyed, why should the same power be limited in its exercise, and myriads of years demanded for the preparation of a home for man?'[32].

Strangest of all, and quite at odds with the opinions of other Evangelical scientists, Brewster speculated that the occupancy of the earth by intelligent beings might have been much longer than was generally believed. Men might have existed in earlier geological epochs but not have left fossil remains, and there might even have been creations pre-dating the primitive azoic formations:

'Another creation may lie beneath—more glorious creatures may be entombed there. The mortal coils of beings more lovely, more pure, more divine than man, may yet read to us the unexpected lesson that we have not been the first, and may not be the last of the intellectual race'[33].

Brewster was like other Evangelical scientists in his zeal to defend the friendship of true science and Christianity. He was equally harsh towards false schemes of reconciliation between science and the Scriptures as he was towards what he considered the infidel and spurious science of the author of *Vestiges*. Yet, like other Evangelical writers, he shifted his ground as to where the boundary lay between legitimate theorising and unwarranted and dangerous speculation. Brewster, however, tended to go to greater extremes than most of his fellow Churchmen. Where others might have been content to engage in apologetics—the process of demonstrating that a particular theory did not conflict with orthodox belief—Brewster could rarely resist the temptation to make additional religious and perhaps even some nationalistic capital out of a particular theory or position. As a result he was sometimes obliged to make a dramatic about-face, notably in the case of the nebular hypothesis.

The reasons for this tendency probably lay partly in Brewster's personality, but they may have stemmed partly from the pressures to which he was subject. More than his fellow Evangelical scientists—men like Fleming and Miller—he was anxious to make his name as a scientific figure outside Scotland and he felt that his attempts were often frustrated by the activities of English, and particularly Cambridge, scientists like William Whewell. I suggest that one factor in his initially enthusiastic advocacy of the nebular hypothesis was his desire to outdo Whewell, who in his *Bridgewater Treatise* had only been lukewarm towards the theory. Similarly his rather amateurish philosophical speculations in the debate on the plurality of worlds may have been partly an attempt to establish a reputation as a metaphysician to make up for what he felt was the less than adequate acknowledgement he had received as a physicist.

Notes and References

1. For accounts of the Ten Years Conflict and the Disruption see Andrew L. Drummond and James Bulloch, *The Scottish Church 1688–1843: The Age of the Moderates* (Edinburgh, 1973) and Hugh Watt, *Thomas Chalmers and the Disruption* (London, 1943).
2. Margaret Maria Gordon, *The Home Life of Sir David Brewster by his daughter* (Edinburgh, 1869), 310–327.
3. [D. Brewster], *An Examination of the Letter addressed to Principal Hill, on the case of Mr Leslie, In a Letter to its Anonymous Author. With remarks on Mr Stewart's Postscript, and Mr Playfair's Pamphlet. By a Calm Observer* (Edinburgh, 1806).
4. E. W. Morse, 'Natural Philosophy, Hypotheses and Impiety: Sir David Brewster confronts the undulatory theory of light', unpublished Ph.D thesis, (University of California, Berkeley, 1972), *passim*.
5. T. Chalmers, *A Series of Discourses on the Christian Revelation, viewed in connection with the Modern Astronomy* (Glasgow, 1817).
6. For more on Miller, see his autobiography, H. Miller, *My Schools and Schoolmasters or, the Story of my Education* (Edinburgh, 1854), and Peter Bayne, *The Life and Letters of Hugh Miller*, 2 vols (London, 1871).
7. For a biography of Fleming see J. Fleming, *The Lithology of Edinburgh; Edited, with a memoir [on Fleming] by the Rev. John Duns, Torphichen* (Edinburgh, 1859).
8. Geologists who favoured such a mode of reconciling the geological record with Genesis were called Scriptural Geologists. See M. Millhauser, 'The Scriptural Geologists, an Episode in the History of Opinion', *Osiris*, 11 (1954), 65–86.
9. See L. E. Page, 'The Rise of the Diluvial Theory in British Geological Thought', unpublished dissertation (University of Oklahoma, 1963).
10. T. Chalmers, review of Cuvier's 'Theory of the Earth', *Edinburgh Christian Instructor*, 8 (1814), 261–274.
11. See L. E. Page, 'Diluvialism and its Critics', in C. J. Schneer (ed.), *Towards a History of Geology* (Cambridge, Mass., 1969), reprinted in C. A. Russell (ed.), *Science and Religious Belief: A Selection of Recent Historical Studies* (London, 1973).
12. [D. Brewster], review of W. Buckland's *Bridgewater Treatise* 'Geology and Mineralogy', *Edinburgh Review*, 65 (1837), 4.
13. C. C. Gillispie, *Genesis and Geology. A Study in the Relations of Scientific Thought, Natural Theology and Social Opinion in Great Britain, 1790–1850* (New York, 1959).
14. T. Chalmers, *On the Power, Wisdom and Goodness of God as Manifested in the Adaptation of External Nature to the Moral and Intellectual Constitution of Man*, 2 vols (London, 1833).
15. [D. Brewster], review of P. M. Roget's Bridgewater Treatise 'On Animal and Vegetable Physiology', *Edinburgh Review*, 60 (1834), 179.
16. For a discussion of the natural theology of the Scottish Evangelicals with specific reference to the work of Chalmers see for instance C. Smith, 'From Design to Dissolution: Thomas Chalmers' Debt to John Robison', *British Journal for the History of Science*, 12 (1979), 59–70. I discuss Evangelical natural theology in greater detail in my forthcoming thesis.
17. Pierre Simon Laplace, *Exposition du Systeme du Monde*, 2 vols. (Paris, an IV [1796]), II, 301–303.
18. William Herschel, 'Astronomical Observations Relating to the Construction of the Heavens, Arranged for the Purpose of a Critical Examination, the Result of which Appears to Throw some New Light upon the Organization of the Celestial Bodies', *Philosophical Transactions of the Royal Society of London*, 101 (1811), 269–345. See also S. Schaffer, 'Herschel in Bedlam: Natural History and Stellar Astronomy', *British Journal for the History of Science*, 13 (1980), 211–239.
19. Nichol's earliest published item on the theory was J. P. N[ichol], 'State of Discovery and Speculation concerning the Nebulae', *London and Westminster Review*, 25 (1836), 390–409 which was followed by a fuller account of the theory in J. P. Nichol, *Views of the Architecture of the Heavens, In a Series of Letters to a Lady* (Edinburgh, 1837).
20. [D. Brewster], review of A. Comte's 'Cours de Philosophie Positive', *Edinburgh Review*, 67 (1838), 276–277.
21. [R. Chambers], *Vestiges of the Natural History of Creation* (London, 1844).
22. [D. Brewster], review of 'Vestiges of the Natural History of Creation', *North British Review*, 3 (1845), 471–472.
23. [D. Brewster], 'The Earl of Rosse's Reflecting Telescopes', *North British Review*, 2 (1845), 175–212.
24. Brewster, *op. cit.* (22), 481.
25. [D. Brewster], 'The Revelations of Astronomy', *North British Review*, 6 (1846), 206–255, contains the admission that he had supported the nebular hypothesis before it became 'the basis of mischievous speculation' (p. 240).
26. D. Brewster, 'The Facts and Fancies of Mr Darwin', *Good Words*, 3 (1862), 3–9.
27. [W. Whewell], *Of the Plurality of Worlds: An Essay*. (London, 1853). For more on the Brewster-Whewell debate see J. H. Brooke, 'Natural Theology and the Plurality of Worlds: Observations on the Brewster-Whewell Debate', *Annals of Science*, 34 (1977), 221–286.
28. D. Brewster, *More Worlds than One. The Creed of the Philosopher and the Hope of the Christian* (London, 1854).
29. *Ibid.*, 256.
30. *Ibid.*, 101.
31. *Ibid.*, 67.
32. [D. Brewster], review of R. Murchison's 'Siluria', *North British Review*, 21 (1854), 506.
33. Brewster, *op. cit.* (28), 52.

Sir David Brewster as an Historian of Science

JOHN R. R. CHRISTIE

For the historian of science, David Brewster occupies a peculiarly privileged position. This is not because he did so much and inhabited such a variety of informative contexts, as the range of papers presented in this volume testifies, but because he is one of the very small number of past scientists who wrote history of science which is still consulted and used as history of science by historians of science. He exists, that is, as both subject and colleague, and that is a rare thing. In his numerous writings, Brewster would often touch upon the history of science, particularly in his reviews, but I wish to concentrate on his two principle historical works, the *Martyrs of Science* published in 1841[1], and the *Memoirs of the Life, Writings and Discoveries of Sir Isaac Newton*, published in 1855, the work on which his reputation as historian of science rests[2].

The reader of *Martyrs of Science*, three short biographies of Galileo, Tycho and Kepler, will be struck, on completing it, by the fact that there are no martyrs in the book. The title indeed was a source of amusement to Brewster's family. 'The significance and quaintness of the title excited much pleasantry', his daughter wrote, 'and long formed an element in the pleasant household raillery which my father was so pre-eminently good-humoured in sustaining and enjoying'[3]. The title to my mind has a double significance, which involves the presence of one sort of martyrdom and the absence of another. Two of his subjects, Tycho and Kepler, suffered a form of martyrdom to which Brewster himself was peculiarly prone, the civil martyrdom inflicted upon men of science by governments, kings and emperors, who for much of history had consistently failed to provide support for men of science and to grant them the recognition, status and state salary they deserved. He detailed the way in which Kepler's pension as Imperial Mathematician was let fall in arrears by the Emperor Rudolph, and commended Kepler's good sense in having declined Ambassador Henry Wotton's invitation to come to England:

'The generous hearts of individual Englishmen, indeed, are always open to the claims of intellectual pre-eminence, and ever ready to welcome the stranger whom it adorns; but through the frozen life-blood of a British minister such sympathies have seldom vibrated'[4].

By refusing Sir Henry, 'the sacred name of Kepler was thus withheld from the long list of distinguished characters whom England has starved and dishonoured'[5]. It does not do to make too light of such assertions, found elsewhere in Brewster's history. They form his mode of identification with scientists of the past, in terms of poverty and civil suffering. And in so far as he had a sense of science's possessing a political history, it was precisely in this negative form of English political negligence, complemented by the positive contrast of France, which had consistently given its scientists the support they required and deserved.

The absence of martyrdom I have mentioned was Galileo's, the biography which starts *Martyrs of Science*. Here Brewster produced a highly interesting and significant inversion of the conventional view of Galileo's trial at the hands of the Inquisition. While deploring Galileo's undoubted persecution, Brewster actually made a brief *apologia* for the Inquisition's attitude[6], and reserved the weight of a genuine fury for Galileo himself. His fury stemmed from the following view of the case. He saw Galileo's career as one openly seeking, and naturally culminating in, martyrdom. But mere persecution did not constitute martyrdom. Galileo failed to take his chance of becoming a genuine martyr by denying the motion of the earth. Brewster found this unintelligible and unforgiveable:

'But what excuse can we devise for the humiliating confession and abjuration of Galileo? Why did this

Fig. 7. Sir David Brewster by Andrew Robertson. Miniature. *(Messrs. Christie, London).*

master-spirit of the age—this high-priest of the stars—this representative of science—this hoary sage, whose career of glory was near its consummation—why did he reject the crown of martyrdom which he had himself coveted...?'[7]

Brewster returned to the case of Galileo in his biography of Newton, and there conjured the following scene:

'—had Galileo added the courage of the martyr to the wisdom of the sage,—had he carried the glance of his eye round the circle of his judges, and with uplifted hands called upon the living God to witness the truth and immutability of his opinions, he might have disarmed the bigotry of his enemies, and science would have achieved a memorable triumph'[8].

The scene is conjured in dramatically pictorial terms, and constitutes not just a comment on what should have happened, but a wish-fulfilment of Brewster's, a passionate corrective to history as it happened.

The fact that Brewster could neither understand nor forgive Galileo his lack of martyrdom makes a sombre point. *Martyrs of Science* may have been a joke to Brewster's family; it may be one to us; but martyrology was a subject of deep seriousness to Brewster. What Brewster implied was that adherence to science was a faith unto death which may not be evaded should circumstances so conspire. You have here, I suggest, the fervour of a Scottish Evangelical wedded to, and largely indistinguishable from, an intense commitment to science and its truths, and this produces an explosive historical judgement.

Before leaving the *Martyrs of Science*, I would draw particular attention to one other section as having especial significance for a problem Brewster was later to encounter in his work on Newton. Brewster made mention of Tycho's belief in astrology and alchemy, as something the historian, however unwillingly and painfully, is required to confront. His framing of the problem was as follows:

'Admitting then, as we must do, that Tycho was not only a professed alchemist, but that he was practically occupied with its pursuits, and continually misled by its delusions, it may not be uninteresting to the reader to consider how far a belief in alchemy, and a practice of its arts, have a foundation in the weakness of human nature; and to what extent they are compatible with the piety and elevated moral feeling by which our author was distinguished'[9].

The problem is posed in moral terms, of human weakness and its apparent compatibility with an otherwise moral character. Brewster has a three part answer, involving an epistemological, a social and a moral component. If we had more knowledge of ancient alchemical beliefs, they would be genetically explicable by 'tracing them ... to the ordinary principles by which the human mind is in every age influenced and directed'[10]—that is, by referring them to the epistemological principles of knowledge and belief analysed by such compatriots as Hume, Smith, Reid and Stewart. If you add to that the fact that older forms of society possessed no system of restraining extravagant beliefs, then one can understand how even the cleverest of minds 'necessarily reflect the peculiarities of the age in which they lived'[11]. In the case of more recent alchemy, Brewster proposed, one can at least see some sort of evidential ground for alchemical belief in the observation of common chemical transformations and extractions. The ardent pursuit of alchemy, finally, Brewster explained by ordinary human corruption—ambition, self-interest, the pursuit of wealth[12].

If *Martyrs of Science*, a relatively short and unresearched work, was something of a pot-boiler for Brewster, the *Memoirs of Sir Isaac Newton* was something else again, a classically monumental Victorian biography, deeply researched, and motivated by strong and specific aims. These aims were originally vindicatory in nature, something one quickly gathers on reading it, and as is confirmed by Brewster's daughter—'for twenty years [he] made it one of his objects to search out every proof and evidence by which he could defend Newton from the charges against his sanity, his probity, and his justice ...'[13]. The chief calumniator was Biot, who had published a manuscript of Huygens, which suggested that Newton had become insane in 1692, and that the consequent 'derangement of intellect' suffered by Newton accounted for the lack of any further important new work by him. Biot had also written that Newton's conduct towards John Flamsteed was due to his derangement[14]. Baily too, in his *Life of Flamsteed*, had, according to Brewster, unfairly reported Newton's conduct towards the Astronomer-Royal[15].

Brewster's wish to vindicate Newton's reputation, coupled with the success of his earlier short *Life of Sir Isaac Newton*[16], were then the origins of his grand biographical project. His ambition was further solidified when in 1837 he gained access to the Portsmouth Collection of Newton manuscripts, where he hoped to find material which would enable him to establish a documentary basis for Newton's purity of reputation[17]. Brewster's interaction with the Portsmouth Collection provides the key to understanding the *Memoirs of Sir Isaac Newton*, and the finally incoherent image of Newton embodied in that work. It is a story of expectations initially fulfilled but ultimately subverted.

Before explaining that comment, let me first try to convey an impression of the work as a whole, its structure, style and flavour. The basic pattern Brewster set out to employ was one of chronological narrative, following Newton through early life, education, scientific research and publication, and public career. This pattern holds partially for volume one, which takes us up to the publication of the *Principia*. Yet even in volume one, a subsidiary pattern begins to impose itself, for Brewster was unable to resist the opportunity to fill in portions of the future history of some of Newton's concerns, either further on in Newton's life than Brewster had chronologically reached, or even more strikingly, up into Brewster's own times. On what occasions does this occur? Most obviously, when Newton's interests coincide with Brewster's obsessions. The four chief of these were optical instrumentation, the theory of colour, the physical nature of light and the reform of English scientific institutions. Thus a discussion of Newton's telescope construction takes off into a narrative of telescope construction and telescopic discovery, culminating with a description of the Earl of Rosse's recent gigantic six foot reflector[18]. On the nature of light the track is followed up to Thomas Young, and Brewster used the chapter as an historical vindication of the corpuscular emission theory of light, attempting to show that Newton's commitment to the optical ether was hypothetical and illustrative only, and that Newton's most basic and consistent position favoured the emission theory[20]. On the reform of English scientific institutions, Brewster was delighted to discover a proposal of Newton's advocating that the Royal Society be staffed by properly salaried pensioners, on the French model, and confessed to having gone so far as to show Newton's scheme to Sir Robert Peel; the result of which act, according to Brewster, was the establishment of the Museum of Practical Geology[21]. The chapter ends with a notable piece of self-plagiarism, Brewster reprinting word for word paragraphs of an earlier *North British Review* article, advocating a National Institute of the Sciences and Arts[22]. The result of all this is to impose upon the chronological narrative a dynamic oscillating structure, the reader being switched back and forth between past and present. Further, a kind of egotistic teleology governs the movement of this pendulum: frequently, Newtonian history culminates in contemporary Brewsterian action, history connects Brewster closely with his biographical subject. The particular strength of volume one lies in its treatment of Newton's optical work, where Brewster was able not just to bring his own general optical expertise to bear, but his understanding of experimental design, and above all his awareness of the importance of the material instrumental basis of scientific practice[23]: something largely absent from the mentalist pages of Whewell[24], and from many pages written nowadays.

If we turn next to volume two, we see how the already vibrating structure of Brewster's biography starts to fall apart. Volume two was to have been the vindicating volume, and with regard to the specific nuisances of Biot and Baily, Brewster was successful. He had discovered additional material, and with the aid of hitherto unknown letters of Newton to Flamsteed, felt able to vindicate Newton's conduct[25]. By demonstrating too how Newton's mind lost none of its power after 1687, Biot's slur is refuted[26]. In so far as it is possible to produce an *apologia pro Newtoni* on these questions, Brewster did so. But although this original ambition was successfully concluded, closer acquaintance with the new manuscripts contained a drastic series of shocking surprises for Brewster. He perused some of these manuscripts, one imagines, with mounting horror as to what they revealed: namely, that Newton was a persistent devotee of alchemical literature and experiment, and that Newton's religious beliefs were of a deeply unorthodox kind.

With regard to Newton's alchemy, Brewster had, early in volume one, dismissed an already extant alchemical letter of Newton's, dating from 1669. This had been treated as an isolated lapse of a 'young philosopher', not worth lingering over[27]. Now he had to face the problem of Newton's abiding alchemical interests, interests seen not only in experimentation but in much reading and annotation of notorious mystics and alchemists such as Jacob Behmen and Basil Valentine. The best he could manage was a remobilisation, in more or less identical wording, of his explanation of alchemy in Tycho: the taste for alchemy was prevalent in Newton's time; there was an empirical ground, if no thorough warrant, for belief in transmutation, and in any case, neither Newton, nor Boyle, nor Locke can be identified with the imposters and charlatans, because they were not motivated by the love of wealth or fame, but by the quest for truth—were not, therefore, alchemists in Brewster's sense of the word[28]. The distinction is one of intentionality, not object or method of study. Scientists seek truth, alchemists seek wealth and fame. All this reads as rather lame pleading, and indeed Brewster finally admitted defeat:

> 'In so far as Newton's inquiries were limited to the transmutation and multiplication of metals, and even to the discovery of the universal tincture, we may find some apology for his researches; but we cannot understand how a mind of such power, and so nobly occupied with abstractions of geometry, and the study of the material world, could stoop to be even the copyist of the most contemptible alchemical poetry, and the annotator of a work, the obvious production of a fool and a knave'[29].

Brewster could not understand, as he could not understand Galileo's refusal of martyrdom.

The discovery of the true nature of Newton's religious opinions had an even more damaging effect upon Brewster. Perhaps what attracted Brewster most to Newton was Newton's obvious and strong religious piety. Thus to learn of Newton's anti-Trinitarianism must have been profoundly disconcerting. Once again Brewster bit the bullet and produced a brilliant, if impossibly tendentious reading of the evidence to show that there was no conclusively unequivocal utterance of Newton's which denied the consubstantiality of Christ[30]. But Brewster could not even convince himself on this point, and admitted, actually in the preface to volume one, that Newton's opinions fell far short of orthodoxy: 'Though adverse to my own, and I believe the opinions of those to whom his memory is dearest, I did not feel myself justified . . . to conceal from the public that which they have long suspected, and must have sooner or later known'[31].

Now, to understand the impact of this upon the biography, a grasp of the imagery and rhetorical vehemence which Brewster had hitherto used in his presentation of Newton is needed. Newton had been characterised by a predominantly religious set of terms. Newton was more than a man of impeccable and universal genius. Science itself, and the realm of external nature in one of Brewster's favourite images, were Temples, and Newton was the High Priest in that Temple. At crucial points, for example the discovery of the inverse square ratio, this imagery is invoked: 'To have been the chosen sage summoned to the study . . . the high priest in the temple of boundless space . . . that sage, that High Priest was Newton'[32]. Brewster had discovered, to put it bluntly, that Newton was an alchemist and a heretic; or to put it figuratively, that the High Priest was sometimes a magician and something of an unbeliever. Retrospectively, then, the whole rhetorical drive of the biography, the sacred image of Newton as High Priest, is nullified. Unable to understand and sympathise with these latter findings, Brewster could not begin to integrate them biographically, and the work unravels into a generalised statement of right scientific method, tailing off finally into a list of Newtonian portraits, poems, inscriptions.

Brewster had fallen victim to the irony which conventionally overtakes the passionately expectant historical researcher. But history, one must add, regularly ignores irony as well as creating it. I have been analysing, after all, the standard biography of Newton which has lasted up until, to be precise, last year, which saw the publication of Professor Westfall's new biography[33]. Despite the structural weaknesses I have indicated, Brewster's *Newton* has done valued service for a century and a quarter. It has done so firstly and simply because it possesses the Victorian biographical virtue of inclusiveness. In addition, there are the strengths I have already mentioned in the historical analysis of optics. I would however claim more for it than might be implied by the designation 'standard biography'—never a description which excites. Brewster, in producing a *summa Newtoniana* for his own time, also set an agenda for Newtonian scholarship. This can be seen not merely in the obvious givens of Newton's life, the optics, the calculus, universal gravitation, the conflict with Leibnitz. Beyond these, the biographical problems of Newton's treatment of Flamsteed, his nervous breakdown, the intellectual challenge of integrating the arcana of Newton's religion and alchemy with his better known achievements, all were first comprehensively brought together within the covers of Brewster's book. It was he too who thoroughly dualised Newton, into scientist and alchemist, the rational and the unintelligible. And it is no exaggeration to say that while the answers produced by modern scholarship differ from Brewster's, that agenda of problems remains substantially the same. Humpty Dumpty is not yet put back together again, which is to say that our own, contemporary historiography still fails, in crucial terms, to advance much beyond Brewster's[34].

Notes and References
1. David Brewster, *The Martyrs of Science: Lives of Galileo, Tycho Brahe, and Kepler* (Edinburgh, 1841).
2. David Brewster, *Memoirs of the Life, Writings and Discoveries of of Sir Isaac Newton*, 2 vols (Edinburgh, 1855).
3. Maria Gordon, *The Home Life of Sir David Brewster* (2nd ed., Edinburgh, 1870), 171.
4. Brewster, *op. cit.* (1), 225.
5. *Ibid.*, 226.
6. *Ibid.*, 80–87.
7. *Ibid.*, 87.
8. Brewster, *op. cit.* (2), I, 279.
9. Brewster, *op. cit.* (1), 173.
10. *Ibid.*, 175.
11. *Ibid.*
12. *Ibid.*, 176.
13. Gordon, *op. cit.* (3), 260.
14. Brewster, *op. cit.* (2), II, 131–134.
15. *Ibid.*, 160.
16. Brewster, *Life of Sir Isaac Newton* (Edinburgh, 1831).
17. Brewster, op. cit. (2), I, vii–xii.
18. *Ibid.*, 62–65.
19. *Ibid.*, 182–192.
20. *Ibid.*, 147–150.
21. *Ibid.*, 102–108.
22. *Ibid.*; *cf.* Gordon, *op. cit.* (3), 149–150.
23. Brewster, *op. cit.* (2), chs. III–X, *passim*.
24. William Whewell, *History of the Inductive Sciences*, 3 vols (London, 1837). Whewell's historiography is notably idealist in character, in contrast with much of Brewster's historiographic emphasis.
25. Brewster, *op. cit.* (2), II, 161–186.
26. *Ibid.*, 131–156.
27. *Ibid.*, I, 35.
28. *Ibid.*, II, 372–374.
29. *Ibid.*, 374–375.
30. *Ibid.*, 337–354.
31. *Ibid.*, I, xv.
32. *Ibid.*, I, 319.
33. Richard Westfall, *Never at Rest. A Biography of Isaac Newton* (Cambridge, 1980).
34. This is not to say that laudable and at times successful attempts to move Newtonian interpretation beyond the limits of what Brewster found unintelligible have not been made; see, notably, B. J. T. Dobbs, *The Foundations of Newton's Alchemy* (Cambridge, 1975). It is rather to say that Brewsterian categories and dichotomies still infect the very framing of the questions we pose, and to that considerable extent have not been transcended.

THE **PATENT KALEIDOSCOPES**, in various forms, and with all the latest improvements, are made and sold,

IN LONDON,

By Messrs P. & G. DOLLOND, St Paul's Church-Yard; Messrs W. & S. JONES, Holborn; Mr R. B. BATE, Poultry; Messrs THOMAS HARRIS & SON, Great Russel Street; Mr BANCKS, Strand; Messrs WM. & THOS. GILBERT, Leadenhall Street; Mr BERGE, Piccadilly; Mr THOMAS JONES, Cockspur Street; Mr BLUNT, Cornhill; Mr SCHMALCALDER, Strand; Messrs WATKINS & HILL, Charing Cross; Mr SMITH, Royal Exchange; and Messrs SPENCER, BROWNING, & RUST, Wapping.

AT BIRMINGHAM,

By Mr PHILIP CARPENTER.

AT BRISTOL,

By Mr C. BEILBY.

AT LIVERPOOL,

By Messrs EGERTON SMITH, & Co.

AND

AT EDINBURGH,

By Mr JOHN RUTHVEN.

Brewster and Scientific Instruments

A. D. MORRISON-LOW

There are very few surviving objects which can be associated with Sir David Brewster, and particularly few scientific instruments. His daughter Mrs. Gordon explains at the end of her biography of her father that by his own wish his books, pictures, instruments and papers were taken to the family home near Kingussie in Inverness-shire[1]. In 1903 there was a disastrous fire there, and although the house has since been rebuilt, little connected with Brewster survives.

Brewster's interest in optical instruments can be divided into four main parts. Firstly, there were the instruments which he improved, in particular those described in his early papers and in his 1813 *Treatise on New Philosophical Instruments*. These ranged from achromatic eyepieces through micrometers to telescopes, and included improvements to microscopes, clinometers, protractors and goniometers. Secondly, there were optical instruments which he actually invented. The most famous of these were the kaleidoscope and the lenticular stereoscope; but there were others which he designed (even if some were not constructed) such as the teinometer, for ascertaining the elasticity of bodies[2], and the lithoscope, which measured refractive indices of crystal materials[3]. Thirdly, his interest in optical practicalities, together with his fame as the leading experimental physicist of his day, helped him with his mission to advance the cause of science in areas such as patent reform[4], and involved him in jury work at the Great Exhibition, where he saw the British instrument trade in direct comparison to its Continental counterpart and found it lacking[5]. This in turn was linked with the fourth strand of his interest which was actively to encourage and patronise various philosophical and mathematical instrument-makers throughout the British Isles.

The main difficulty with his fourfold interest in optics and particularly in this last mentioned area of it, was Brewster's own lack of hard cash. Brewster was working outside the security of the professional scientific community—such as it was—which was largely located within the universities. He himself was unable to pursue a university teaching career due to a nervous inability to speak in public[6]. Lacking private means, he was thus forced to turn for a means of support for himself and his growing family to the precarious and insecure life of a journalist, with debts and lawsuits threatening him with apparent disaster almost constantly. It was no wonder that he burned with injustice at his situation, increasingly identifying the cause of science with ideas of religion and progress, seen from the example of his own career. 'No such favourite anywhere as Sir David Brewster, except at home or with anyone engaged with him in business; nobody ever had dealings with him and escaped a quarrel'[7]. Thus Elizabeth Grant of Rothiemurchus wrote of him in later life.

His interest in scientific instruments, together with his interest in science, began early in life. He was encouraged by various childhood friends, of whom Mrs. Gordon mentions at least half a dozen, singling out the greatest influence as one James Veitch of Inchbonny, who was ten years older than Brewster. By the time Brewster was ten he had constructed a telescope under Veitch's guidance, and although two years later he left Jedburgh for a theological training at Edinburgh University, correspondence between the two continued for many years. Veitch was a self-educated astronomer and mathematician, and although during his life-time he achieved local acclaim for his scientific and technical skills, he never moved from his native Jedburgh, apparently lacking the ambition and desire for fame, and being perfectly happy as he was. Brewster wrote to him about his scientific pursuits while following his divinity course; he discussed instrumentation, criticised various Edinburgh opticians and later asked Veitch for various practical commissions[8].

Fig. 8. Trade card of sellers of the kaleidoscope. *(Science Museum, London)*.

Brewster's first book, the *Treatise on New Philosophical Instruments* was published in 1813, by which time he was thirty-two. It was brimming with ideas for improvements and new instruments, and with tables of refractive indices and dispersive powers, together with Brewster's early observations on polarised light. It thus contains a fairly comprehensive introduction to those subjects with which Brewster remained involved throughout his long and busy life. It is in this book that Brewster mentions various instrument-makers to whom he went for his own research apparatus, as distinct from his popular inventions. Among them he mentions William Harris of London, constructing a goniometer to measure the angles of crystals in February 1809, and Alexander Adie, the Edinburgh optician, modifying it in April[9].

Alexander Adie's uncle John Miller, with whom he was in partnership from 1804 until 1822, had learned his instrument making trade in London under the fashionable George Adams, mathematical instrument-maker to George III. Adie himself went into partnership with his second son John in 1835, but ceased to be active in the business about 1837; he died in 1859. His connection with Brewster appears to have continued from the early days, and he was one of four Scottish opticians—James Veitch was another—commissioned by Brewster to manufacture jewel lenses for microscopes in the mid-1820s[10].

Brewster suggested in the *Treatise on New Philosophical Instruments* that in the search for greater image clarity in the microscope, diamonds might be used as lenses instead of glass, as they combine a high refractive power with a low dispersive power. His idea rested for a decade or so, during which time Adie contributed articles on meteorology to Brewster's *Edinburgh Journal of Science*, having invented and patented an oil barometer for use at sea, called the sympiesometer, in 1818[11]. The instruments he produced at this time were all of a high standard, and one interesting survival is a reflecting microscope, modelled on the then recently introduced instruments of G. B. Amici of Modena, showing that Adie, like Brewster, was actively involved in an area that had become a pressing concern amongst progressive instrumentalists[12].

Brewster's pursual of his idea of using precious stones instead of glass lenses was initially discouraged by the seeming impracticality of the task, although the Edinburgh optician Peter Hill evidently made him one lens of garnet and one of ruby[13]. However, in the mid 1820s two London optical workers, C. R. Goring and Andrew Pritchard successfully made and sold microscopes containing diamond and sapphire lenses. Brewster himself employed four opticians to make jewel lenses[14], but of those only Adie seems to have prepared any for sale. These do not appear to have been a commercial success, being expensive and difficult to produce, and as has been pointed out elsewhere[15], the jewel lens was soon technically superceded by the cheaper Wollaston doublet and the newly improved achromatic compound microscope, which was to be proved a major research tool in 19th century science.

Despite this setback, for many years Brewster continued to claim the superiority of jewel lenses over glass because of their durability, which compared well with more unstable flint glasses[16]; this characteristic doggedness was another facet to his frequent claims of priority of invention. A later example of such a claim was that of the so-called Coddington lens. One of Brewster's interests was that of reducing the degrading effects of aberrations on the image formed in both the telescope and the microscope. In an article he wrote for the *Encyclopaedia Britannica* he described his so-called 'Grooved Sphere':

> 'This lens derives its name from its having a deep groove cut round it in the plane of a great circle perpendicular to the axis of vision. Sir David Brewster was led to its construction by the doublet of Dr. Wollaston, . . . It consists of a spherical lens or sphere, with a deep concave groove cut round it, so as to cut off the marginal pencils and thus give a wider field and a more perfect image.'

This is followed by a furious footnote:

> 'This lens, which has very incorrectly been called the "Coddington lens" was not invented by Mr. Coddington, nor was its invention ever claimed by him. It was invented by Sir David Brewster, and constructed in Edinburgh of glass and garnet long before 1820, and an account of it was published in the *Edinburgh Philosophical Journal*, April–July 1820. Mr. Coddington takes no notice of it in his *Treatise on Optics*, published in 1823, but he mentions it without any name in his Treatise *On the Reflection and Refraction of Light*, published in 1829. Mr. Coddington we have reason to believe, got a grooved sphere made by Mr. Carey or some other London optician who, supposing it to be new, gave it a name to which it was not entitled. Had Mr. Coddington been alive, he would have been one of the first to give it its true name'[17].

This sort of emphatic defence of his own work, often justified, is reiterated time and again, notably over the polyzonal lenses in British lighthouses[18] and of course, the disaster of his most famous invention the kaleidoscope.

It is now impossible to estimate whether Brewster 'rediscovered' or 'invented' the kaleidoscope, relying as it did on principles which had been known (but had remained unexploited) since antiquity; or whether he realized that merely patenting his device under a name coined by himself did not make his legal rights to the invention secure if the patent was breached. Brewster 'discovered' the principle of the kaleidoscope while experimenting with the polarisation of light in 1814 and 1815, and although its proud inventor possibly overrated its uses, the instrument undoubtedly has brought joy to millions. He wrote:

'... it was impossible not to perceive that it would prove of the highest service in all the ornamental arts, and would, at the same time, become a popular instrument for the purpose of rational amusement. With these views, I thought it advisable to secure the exclusive property of it by a Patent...'[19].

This application, in 1817, was made under an accumulation of legal bureaucracy which had been piling up since medieval times: it was a procedure which was both time consuming and expensive, costing a minimum of £310, with major drawbacks affecting the inventor. Among these were the risk that the invention might be pre-empted by premature disclosure during the patenting procedure, and also the fact that the legal proceedings for enforcing the patent rarely took place in a single court, and could take years and prove extremely expensive[20].

Like some early nineteenth century Rubik's Cube, Dr. Brewster's kaleidoscope became an overnight bestseller, with an estimated two hundred thousand instruments sold in London and Paris during three months. 'Out of this immense number', wrote the understandably peeved Brewster,

'there is perhaps not one thousand constructed upon scientific principles and capable of giving anything like the correct idea of the power of the Kaleidoscope; and of the millions who have witnessed its effects, there is perhaps not an hundred who have any idea of the principles upon which it is constructed, who are capable of distinguishing the spurious from the real instrument, or who have sufficient knowledge of its principles for applying it to the numerous branches of the useful and ornamental arts'[21].

So, one wonders, what went wrong? Elsewhere Brewster wrote:

'After the patent was signed, and the instrument in a state of forwardness, the gentleman who was employed to manufacture them under the patent, carried a kaleidoscope to show the principal London opticians, for the purpose of taking orders from them. These gentlemen naturally made one for their own use, and for the amusement of their friends; and the character of the instrument being thus made public, the tinmen and the glaziers began to manufacture the detached parts of it, in order to evade the patent; while others manufactured and sold the instrument complete, without being aware that the exclusive property of it had been secured by a patent'[22].

Who was the 'gentleman' who so mismanaged Brewster's brainchild? A trade-card (Fig. 8) giving a list of authorised kaleidoscope retailers who presumably had Brewster's blessing exists in the Science Museum collection[23]. It can be dated from the addresses of their businesses to between 1817 and 1820. The patent itself dates from 30 August 1817 but it is possible that this card dates from after the 'gentleman's' visit to London and thus may illustrate Brewster's efforts to make at least some money out of the whole bungled affair by licencing the retailers. The instruments, however, seem to tell an independent story. The early examples appear to divide into two types: the polyangular kaleidoscope, and the telescopic model. The polyangular seems to have been made exclusively by R. B. Bate, who is named on the trade card, and his item would be relatively expensive to produce. None are known to have a serial number over 60, which suggests that either Bate was slow to produce them, or that they did not sell easily[24]. The unnumbered telescopic model is more common. Philip Carpenter, mentioned on the trade card as 'at Birmingham', marked his instruments on the eyepiece 'P. Carpenter Sole Maker' and on the other end 'Dr. Brewster's Patent'; although later examples still say 'P. Carpenter Sole Maker' they no longer say 'Dr. Brewster's Patent'. The Carpenter kaleidoscopes are very similar to the Dollond ones, and it is known that Carpenter supplied Dollond with telescopes: so probably Carpenter's retailing trade was large[25]. It was not large enough, however, to cope with the great kaleidoscope boom, for Brewster wrote to his wife from Sheffield on 17 May 1818 about his visit to

'... Cam and Cutt, who have undertaken to manufacture the kaleidoscope for Mr. Ruthven [of Edinburgh]. They have agreed to make and sell the instruments under my patent on the same terms as Mr. Carpenter, provided I get his permission to allow them to be employed. This I must do, as he cannot possibly supply the demand'[26].

So it wasn't Carpenter who bungled the affair; nor Bate, of whose model Brewster was very proud. In fact Brewster wrote to J. D. Forbes in 1831:

'Among the Opticians in London I would advise you to call on Mr. R. B. Bate in the Poultry, one of the ablest and certainly one of the *best* men among them. He has a conscience as well as a head which cannot be said of them all'[27].

It can be conjectured that the 'gentleman' was William Harris, to whom Brewster had gone in 1809 for a goniometer, and with whom he took out a patent for a micrometer telescope in 1811. William Harris did produce a kaleidoscope—there is an example in the Science Museum—but despite his previous associations with Brewster, he was not included on the kaleidoscope trade card, nor apparently was he patronised by Brewster again, although the firm continued until 1848[28].

However, last word on the kaleidoscope for the moment should come from Elizabeth Grant, who remembered 1818 with less bitterness than the Brewsters:

'At the Brewsters they ... entertained, ... not in the flat where we first found them, but in their own house in Athole [actually Coates] Crescent newly built out of the profits of the Kaleidoscope, a toy that was ridiculously the rage from its humble beginning in the tin tube with a perforated card in the end, to the fine brass instrument set on a stand, that was quite an ornament to the drawing-room. Had Sir David managed matters well, this would have turned out quite a fortune to him: he missed the moment and only made a few thousand pounds; still, they gave him ease, and that was a blessing'[29].

Brewster's other major successful invention was the lenticular stereoscope, announced in 1849, which was another popular scientific toy, only outsold during the nineteenth century by the kaleidoscope. Brewster's model was inspired by an earlier form of the instrument, invented by Sir Charles Wheatstone between 1830 and 1832, but took advantage of the discovery of photography in 1839, whereas Wheatstone had not pursued his initial attempt to use this. Brewster, of course, vigorously denied the similarity of the two instruments, and wrote a defence strongly biased from his point of view in *The Stereoscope*, published in 1856. He and Wheatstone also conducted a prolonged and verbose disputation over the origins of the stereoscope in the correspondence columns of *The Times* that same year[30].

The lenticular stereoscope was perhaps Sir David's most famous invention connected with photography, a subject with which he had been involved since its English beginnings in the 1830s. The stereoscope gives an illusion of a three dimensional scene from two slightly different flat pictures which are viewed through the apparatus so that each eye sees only one picture. It is the three inches or so between the human eyes which allows the brain to use these different images to determine distance and relief. Wheatstone's model did it with mirrors: Brewster's with lenses. Brewster's paper 'Description of several New and Simple Stereoscopes for exhibiting, as Solids, one or more representations of them on a Plane' was read before the Royal Scottish Society of Arts on 26 March 1849 and was followed by his 'Account of a Binocular Camera, and of a Method of obtaining Drawings of Full Length and Colossal Statues, and Of Living Bodies, which can be exhibited as Solids by the Stereoscope'.

Curiously, Brewster does not seem to have gone to the same instrument-maker for both items. Taking his stereoscope first: 'Stereoscopes made by Mr. Loudon, optician, Dundee, ... were sent to several of the nobility in London, and to other places in England' wrote Brewster[32]. George Lowdon of Dundee began making scientific instruments as a business in May 1849, then aged 24, and recounts the story in his autobiography:

'Fortunately for me, at the end of 1849 I got acquainted with that nobleman so well and favourably known to all Dundonians, George, Lord Kinnaird, and through him was introduced to many of the savants who were entertained by his Lordship at Rossie Priory. Amongst these was Sir David Brewster, who had at this period (1849) invented his stereoscope, and I got the making of the first one, and the sending of copies of it to many scientific men all over Europe. Later on I also improved on them, and made a great number for many years afterwards. The fault of Brewster's stereoscope was that the lenses were too small, being in fact, only the two halves of a spectacle glass. This did not suit every eye, and in experimenting I discovered that larger lenses were an advantage. I pointed this out to Sir David, but he was wedded to his opinion, and as I feared the idea might be taken up by another, I took out a patent for my improvement—which experience has since amply justified—but my action was, unfortunately, resented by Sir David, and gave rise to considerable friction, for which I did not consider I was to blame, seeing I had pointed out the improvement and he had refused it'[33].

Thus was another optician discarded. Brewster was unable to persuade any British makers to construct his stereoscope, and so took one of Lowdon's examples to Paris with him in the spring of 1850. There he showed it to M. Soleil and his son-in-law M. Duboscq, probably because of their connection in producing excellent polarisation apparatus with which he would have been familiar. They started to produce the viewer, with daguerreotype images made with a rudimentary stereo camera, and displayed one at the Great Exhibition of 1851. 'The stereoscope attracted the particular notice of the Queen, and M. Soleil executed a beautiful instrument which was presented to Her Majesty in his name by Sir David Brewster'[34]. With the Royal seal of approval, it could not fail commercially. As Brewster does not complain, one can only conclude that he must have made some money out of it, or at least lost less than he expected.

However, Lowdon does not seem to have constructed Brewster's binocular camera, though from 1849 he had

'made many cameras and given instruction to a large number of professionals and amateurs—notably to the late artist-photographer, Mr. G. W. Wilson of Aberdeen. I sold him a camera in 1853, and gave him his first lesson'[35].

From Brewster's own account it would appear that by 1851 a 'Mr. Slater of Euston Square has already constructed several of these binocular cameras, which have been sent to America'[36]. However, the first successful binocular camera to be patented and sold commercially in England was made by J. B. Dancer, of Liverpool. In his autobiography, Dancer claims that Brewster showed him the lenticular stereoscope, which inspired Dancer to 'construct a convenient form of Camera for photographing the pictures simultaneously'[37]. Brewster took great interest in another Dancer invention, that of microscopically reduced photographs, taking examples abroad with him in 1856, where the idea was impudently patented in France by the photographer P. R. P. Dagron. This microphotographic technique was used during the siege of Paris in 1870 when the Parisians sent messages by carrier pigeon over the German lines to the outside world. Brewster also encouraged Dancer to use this process in the production of micrometers and diffraction gratings for telescopes and microscopes.

There is one further possible instrument maker who may have constructed Brewster's binocular camera for him: Thomas Davidson of Edinburgh, who was active between 1840 and 1853. Davidson contributed many papers on photography and photographic equipment in the early volumes of the *Transactions of the Royal Scottish Society of Arts* and the *British Journal of Photography*. Brewster wrote to W. H. Fox Talbot in October 1840: 'I have got a very fine Camera constructed by Mr Thomas Davison [*sic*] The Royal Exchange Edinr...'[38], and a year later:

> 'Mr Davidson makes the most beautiful Portrait Cameras of which we have two here [i.e. St. Andrews] in constant operation... He is an admirable person of real knowledge, and charges less than half of London artists. I think a Portrait Camera about 12 or 13 inches long with two excellent achromatic lenses w^d cost £5 or £6, but this is more a conjecture than a certainty'[39].

Brewster in fact bought photographic equipment from Davidson in 1841 for £8.10.0. He bought it, among other items, through the University Apparatus Fund, which suggests that the quality of his own scientific equipment may well have suffered from his lack of cash[40].

Brewster's correspondence with the inventor of positive/negative photography, the Wiltshire landowner William Henry Fox Talbot, began in the early 1830s and lasted until his death. Although photography became an absorbing subject for them both, they also discussed other aspects of optical science, among them Brewster's discoveries connected with spectroscopy and Fraunhöfer's absorption lines in the solar spectra and work concerning the polarisation of light. It was this last area that Brewster had first made his reputation among the scientific community of Europe, in the years after his graduation from Edinburgh University in 1800. Early forms of polarising equipment, used by Brewster in Scotland, and Malus and Biot in France, appear to have been fairly rudimentary, initially using polarisation by reflection, where the polariser was a sheet of black glass and the analyser a similar piece of glass or a rhomb of Iceland spar[41]. Next, plates of tourmaline crystal or agate were used, but the great improvement in the design of polarising apparatus came in 1828 with the invention of the Nicol prism by William Nicol, a lecturer on natural philosophy in Edinburgh.

Very little is known about Nicol, who published his discovery in the *Edinburgh New Philosophical Journal*, which by 1828 Robert Jameson was publishing without Brewster's help. Brewster, however, evidently knew Nicol, for he mentions using minerals from Nicol's cabinet in 1825[42] — before the prism was invented — and again in 1853 when he refers to 'the late Mr William Nicol', who had died two years before[43]. Was Brewster, then, jealous of Nicol's invention? The Edinburgh scientific community was numerically small, and it would be reasonable to suppose that two men with common interests in practical optics and crystallography at the same time and in the same place would be bound to know each other, and be familiar with each other's work. But it appears that the Nicol prism article only caught W. H. Fox Talbot's eye through a German journal a few years after its first announcement[44]. Had Brewster managed to avoid telling Talbot about the prism that effectively superceded his own polarisers? Talbot had some made for him, and in 1834 described how they might be incorporated into a compound microscope[45]. In Brewster's article 'On the Microscope' written for the *Encyclopaedia Britannica*, he tells how he (Brewster), used a form of polarising microscope in the early 1820s. Using 'the simple microscope for these purposes, he cemented plates of agate and torumaline with Canada balsam to the plane side of a plano-convex lens'. Brewster credits W. H. Fox Talbot with first using the 'ingenious Nicol prism' with the compound microscope, and suggests that an improvement would be 'to screw the analyser into the lower end of the body of the microscope, immediately behind the object-glass'[46]. In 1839 Brewster mentioned in a letter to Talbot, discussing his failure in certain crystallisation experiments, that

> 'I should have continued to make the attempt, had not a friend who saw Mr Nicol (inventor of the Nicol prism) learned from him that he had experienced the same difficulty, but had a found a way of surmounting it'[47].

This would suggest that during 1839 at least, Brewster and Nicol, although actively working in the same areas, were not communicating with each other about their work. It also lends support to the theory that Nicol, living with his elderly unmarried sister and two servants, was something of a recluse[48]. He was, after all, some years Brewster's senior, and after the invention of his prism had already had a prolonged and bitter argument in print over a method of microscope slide preparations with Henry Witham, author of *Observations on Fossil Vegetables*[49]. No doubt Nicol had no wish to have any dispute with such a well-seasoned adversary as Sir David Brewster.

It can be concluded that in general Brewster's patronage of instrument makers was regarded as a somewhat mixed blessing by those involved. Despite his reputation as the greatest living experimental scientist, he was able to upset and antagonise those very people who might have exploited his ideas and produced from them some form of steady income: 'Sir David Brewster! He lives in St. Andrews and presides over its principal college, yet no-one speaks to him!' wrote Lord Cockburn in 1844.

> 'With a beautiful taste for science, he has a stronger taste for making enemies of friends. Amiable and agreeable in society, try him with a piece of

business, or with opposition, and he is instantly, and obstinately, fractious to the extent of something like insanity. With all arms extended to receive a man of whom they were proud a few years ago, there is scarcely a hand that he can now shake'[50].

Although this contemporary was referring to Brewster's relationship with his university colleagues at St. Andrews, it seems to sum up his conduct towards people in general. His inabilities as a manager and politician helped to make him turn to his pen as a weapon in his vigorous campaign for patent reform and governmental support for science.

Notes and References

1. M. M. Gordon, *The Home Life of Sir David Brewster* (Edinburgh, 1869), 414.
2. D. Brewster, 'On the Laws which Regulate the Distribution of the Polarising Force in Plates, Tubes and Cylinders of Glass, that have received the Polarising Structure', [read 17 June 1816], *Transactions of the Royal Society of Edinburgh*, 8 (1818), 356–371. No examples of this instrument are known.
3. D. Brewster, 'Description of the Lithoscope, an Instrument for Distinguishing Precious Stones and other Bodies', *Transactions of the Royal Society of Edinburgh*, 23 (1864), 419–424. No examples are known to survive, although Brewster states that his own example was made by George Dollond of London.
4. Among other papers are: [D. Brewster], 'The Paris Exposition and the Patent Laws', *North British Review*, 24 (1855), 231–267; [D. Brewster], 'Decline of Science in England and Patent Laws', *Quarterly Review*, 43 (1830), 305–342; D. Brewster, 'Presidential Address', *Report of the British Association for the Advancement of Science . . . 1850* (London, 1851) xxxi–xliv. Also, Gordon, *op. cit.* (1), 207–212.
5. *Exhibition of the Works of Industry of All Nations, 1851. Reports by the Juries on the Subjects in the thirty classes into which the exhibition was divided* (London, 1852) Vol. II Reports–Classes V to XVI, 519–704; David Brewster, *Paris Universal Exhibiton: Report on Certain Optical and other Instruments* (London, 1857); Gordon, *op cit.* (1), 212, 222–225; Elizabeth Gaskell, *Life of Charlotte Brontë* (London, 1975), 463.
6. Gordon, *op. cit.* (1), 56–57.
7. Lady Jane M. Strachey (ed.), *Memoirs of a Highland Lady: The Autobiography of Elizabeth Grant of Rothiemurchus . . . 1797–1830.* (Edinburgh, 1898), 248.
8. National Register of Archives, (Scotland), Survey 337; Veitch as an instrument maker is assessed by T. N. Clarke, *Scottish Scientific Instruments from the Frank Collection in the Royal Scottish Museum* (Edinburgh, *forthcoming*).
9. D. Brewster, *A Treatise on New Philosophical Instruments* (Edinburgh, 1813), 90–91. This instrument has not been traced.
10. G. L'E. Turner, 'The Rise and Fall of the Jewel Microscope 1824–1837', *Journal of the Quekett Microscopical Club*, 31 (1968), 85–94; R. H. Nuttall and A. Frank, 'Jewel lenses—a Historical Curiosity', *New Scientist*, 53 (1972), 92–93; R. H. Nuttall and A. Frank, 'Makers of Jewel Lenses in Scotland in the Early Nineteenth Century', *Annals of Science*, 30 (1973), 407–416.
11. A. Adie, 'Description of the Patent Sympiesometer or New Air Barometer', *Edinburgh Philosophical Journal*, 1 (1819), 54–60; the Adie business is discussed in Clarke, *op. cit.* (8).
12. In the collection of the Royal Scottish Museum, Edinburgh (Inv. RSM TY 1936.144.) Brewster, too, wrote about reflecting microscopes, in *Edinburgh Philosophical Journal*, 8 (1828), 326–327. Another microscope, recently acquired by the Royal Scottish Museum (Inv. RSM TY 1982.96,) and discovered since this paper was read, appears to have been especially commissioned from Adie by the Royal Society of Edinburgh, between the years 1823 and 1829. Brewster was the Society's General Secretary between 1819 and 1828. The instrument is of unusual size, and has an early polarising accessory consisting of a stack of glass plates, as recommended by Brewster in various papers. See A. D. Morrison-Low 'The Origins of the Polarising Microscope: Sir David Brewster *versus* William Nicol' (*forthcoming*).
13. D. Brewster, *A Treatise on Optics* [*Cabinet Encyclopedia*] (London, 1831), 337. Nuttall & Frank, 'Makers of Jewel Lenses . . .' *op. cit.* (10), assume that the garnet lens lens mentioned here is that in the British Museum microscope attributed to Brewster (lens 'R' of Turner's paper, *op. cit.* (10)). See Appendix Ia, item 17.
14. D[avid] B[rewster], 'Microscope' in *Encyclopaedia Britannica*, 21 vols. (8th edition, Edinburgh, 1853–1860), XIX, 765–768.
15. Turner, *op. cit.* (10); Nuttall & Frank, *op cit.* (10); R. H. Nuttall, *Microscopes from the Frank Collection* (Jersey, 1979), 16.
16. [D. Brewster], 'The Microscope and its Revelations', *North British Review*, 25 (1856), 448.
17. Brewster, *op. cit.* (14), 769–770.
18. R. W. Munro, *Scottish Lighthouses* (Stornoway, 1979), 97–108.
19. D. Brewster, *Treatise on the Kaleidoscope* (Edinburgh, 1819), 7.
20. For a history of the patent laws, see Neil Davenport, *The United Kingdom Patent System: a Brief History* (Havant, Hants., 1979).
21. D. Brewster, *op. cit.* (19), 7.
22. [D. Brewster], 'Kaleidoscope' in *Edinburgh Encyclopaedia*, 20 vols (Edinburgh, 1830), XII, 410–412.
23. H. R. Calvert, *Scientific Trade Cards in the Science Museum Collection* (London, 1971), 28, item 223. This list is duplicated at the end of Brewster, *op. cit.* (12) and (19), and is illustrated here as Fig. 8.
24. Bate kaleidoscopes in public collections in the United Kingdom are numbered as follows: Science Museum, London: 45; Whipple Museum of the History of Science: 30 and 40; Royal Scottish Museum: 37. See Appendix 1a, item 4.
25. S. Timmins (ed.), *The Resources, Products and Industrial History of Birmingham* (London, 1866), 534. For more about the history of the firm of Carpenter see R. H. Nuttall, 'Philip Carpenter and the "Microcosm" Exhibition', *Microscopy, 33* (1966), 62–65.
26. Gordon, *op. cit.* (1), 96.
27. Correspondence and papers of James David Forbes (1809–1868) held at the University Library, St. Andrews; that Brewster used Bate as a London agent is supported by a letter in the Royal Scottish Society of Arts archives, held in the National Library of Scotland. Cited in A. D. C. Simpson, 'Brewster's Society of Arts and the Pantograph Dispute' *(forthcoming)*, the letter from Andrew Smith, inventor, to David Brewster, dated 23 June 1821, states: 'I have a letter from My Brother, today saying that he has delivered your letter to Mr. Bate who strongly recommends taking out a patent for England and for this purpose he would agree to advance £100 on the following terms—viz. the Patent to lie in his hands as security for the money laid out. That he shall be allowed to manufacture the whole of the Ins[trument] charging the *nett expence* of workmanship and materials. And to have 25/-6d upon the Same for his outlay and trouble till once we enabled to pay him the cash—with 15/-6d for all that he sells—and should the profits never pay the Patent Mr. B. still has the same claim on us for his money.' Possibly Bate's cautious terms were a direct result of the kaleidoscope experience. I would like to thank Allen Simpson for this reference.
28. Appendix Ia, item 3. E. G. R. Taylor, *The Mathematical Practitioners of Hanoverian England 1714–1840* (Cambridge, 1966), 357, 365, gives outline biographies of both David Brewster and William Harris.

29. Strachey, *op. cit.* (7), 247–248.
30. Brian Bowers, *Sir Charles Wheatstone F.R.S 1802–1875* (London, 1975); R. S. Clay, 'The Stereoscope: Presidential Address', *Transactions of the Optical Society*, *29* (1928), 149–166; A. T. Gill, 'Early Stereoscopes', *Photographic Journal*, *109* (1969), 546–559, 606–614, 641–651.
31. Published in *Transaction of the Royal Scottish Society of Arts*, *3* (1849), 247–259 and 259–264.
32. [D. Brewster], 'Binocular vision and the Stereoscope', *North British Review*, *17* (1852), 176.
33. A. H. Millar, *James Bowman Lindsay and other Pioneers of Invention* (Edinburgh, 1925), 86. Lowdon is assessed as an instrument-maker by Clarke, *op. cit.* (8).
34. Brewster, *op. cit.* (32), 177.
35. Millar, *op. cit.* (33), 93.
36. Brewster, *op. cit.* (32), 181. However, none of these cameras are known to have survived, and Slater is not known to have made any others.
37. 'John Benjamin Dancer F.R.A.S. 1812–1887: an Autobiographical Sketch with some Letters', *Memoirs and Proceedings of the Manchester Literary and Philosophical Society*, *107* (1964–65), 115–142.
38. Science Museum, London: Science Museum Library, Archive Collection (Fox Talbot Papers), Letter dated 23 October 1840. See Appendix Ia, item 22.
39. *Ibid.*, 22 October 1841; other enthusiastic references to Davidson by Brewster are contained in this collection: e.g. 5 October 1840 '... Daguerreotypes executed in Edin. by a Mr Thomas Davison [*sic*] far surpass any done in Paris or London'; 8 November 1840 'An artist in Edin. Mr Thomas Davidson No. 12 Royal Exchange has made very fine Cameras for taking portraits. He is making a very fine one for me; and is an artist of great knowledge and ingenuity,— He has executed Daguerreotypes far surpassing those made by Daguerre. I have two that appear almost miraculous. I have no doubt that his Camera surpasses any other yet made; as he has made a beautiful improvement and in the frame to Portraits'; 28 November 1840 'I expect in a day or two from Mr Davidson a full account of his Process, and I shall let you know the time in which the Picture is formed in The Camera.'
40. D. Smith, 'Inventory of the apparatus in Nat. Philosophy Class of the *United College May 1847*'; the relevant entry in the Ms being 'Paid from app.s fund/to Sir D. Brewster/1839 May 25 4.8.7½/do . . . 16.15.6/1840 July 28 for a ~~telescope~~ microscope 52.8.8/1841 July 9 Davidson a Camera 8.10–/£82.0.9½': Hay Fleming Reference Library, St. Andrews. See introduction to Appendix Ia in this volume for a discussion of Brewster's laboratory and equipment.
41. F. C. Cheshire, 'Polariscopes: a few Typical Forms of Early Instruments in the South Kensington Museum', *Transactions of the Optical Society*, *23* (1921–22), 246–255; F. C. Cheshire, 'The President's Address: The early History of the Polariscope and the Polarizing Microscope', *Journal of the Royal Microscopical Society*, *43* (1923), 1–18.
42. D. Brewster, 'On the Refractive Powers and Other Properties of the two New Fluids in Minerals', *Transactions of the Royal Society of Edinburgh*, *10* (1826), 424 note. For more about the possible relationship and tensions between Brewster and Nicol, see Morrison-Low, *op. cit.* (12).
43. D. Brewster, 'On Cavities in Amber containing Gases and Fluids', *London, Edinburgh and Dublin Philosophical Magazine and Journal of Science*, 4th series *5* (1853), 233–235.
44. [W.] H. F. Talbot, 'Facts relating to Optical Science No. II, 7. On Mr Nicol's Polarizing Eye-piece', *London and Edinburgh Philosophical Magazine and Journal of Science*, 3rd series *4* (1834), 289.
45. [W.] H. F. Talbot, 'Experiments on Light', *London and Edinburgh Philosophical Magazine and Journal of Science*, 3rd series *5* (1834), 321–335.
46. Brewster, *op. cit.* (14), 789.
47. Science Museum, London: Science Museum Library, Archive Collection (Fox Talbot Papers), letter dated 14 March 1839. An edition of the Brewster-Talbot correspondence held in the Science Museum is now in preparation, edited by Joanna M. Lindgren-Harley.
48. Scottish Record Office: William Nicol's Will SC70/4/17.
49. W. Nicol, 'Observations on the Structure of Recent and Fossil Coniferae', *Edinburgh New Philosophical Journal*, *16* (1834), 137–158. An assessment of this dispute is given by N. Higham, *A Very Scientific Gentleman: The Major Achievements of Henry Clifton Sorby* (Oxford, 1963), 35–36 and 51–54.
50. Henry Cockburn, *Circuit Journeys* (Edinburgh, 1888), 234.

Brewster on the Nature of Light

G. N. CANTOR

Of the several hundred books, papers, reviews and encyclopaedia articles written during Brewster's long and industrious career, the subject of optics occurred more frequently than any other. As one biographer noted he was 'distinguished especially for his original discoveries in the science of optics'[1]; indeed, his scientific contemporaries knew him principally as one of the leading, if not the leading, writer and researcher on physical optics. The design of optical instruments, polarisation, double refraction, absorption, spectroscopy, thin plates, phosphorescence, the theory of colours and the structure of the eye—these and numerous related topics were his principal subjects of research. To discuss any one of these adequately would require my allotted space. While other papers in this volume address a few of these topics, the present paper concentrates on just one aspect of Brewster's optical concerns—his views about the nature of light.

This topic provides a theme spanning all seven decades of his adult life. Moreover, it is a particularly revealing topic since the consensus among optical writers changed dramatically over that period[2]. In the 1790s, when Brewster first became interested in optics, almost all writers on natural philosophy followed Newton's suggestions and claimed that light was composed of small particles of matter emitted from luminous sources. Moreover, they accounted for all deviations from rectilinear propagation—as in the phenomena of reflection, refraction and inflection—by short-range *forces* operating at the surfaces of material bodies. Few writers, however, adhered to wave theories of light. By contrast, at the end of Brewster's life this projectile theory was abandoned by the majority of teachers and research physicists. Instead they adhered to a wave theory of light, derived principally from Augustin Fresnel and elaborated by many hands, which was able to explain a vastly-increased range of optical phenomena with impressive accuracy. Thus during his life-time Brewster witnessed a 'revolutionary' change in his favourite subject—from a projectile to a wave theory; indeed, he was particularly active in the 1830s when the 'optical revolution' was at its peak.

Our discussion is intended to make two general points about Brewster. Standard accounts[3] emphasise his great devotion to Newton, on whom Brewster modelled his scientific work, holding steadfastly and uncritically to Newton's projectile theory of light long after his contemporaries had abandoned that theory. Thus by the 1840s he had become an embittered reactionary unwilling and unable to adapt to new, progressive intellectual currents. Moreover, he is often portrayed as a dull fact-gatherer 'filling in the minutiae of the existing system'[4] but unconcerned with theoretical or philosophical issues. By concentrating on Brewster's early scientific career and on his later dispute with the proponents of the wave theory I intend to show that he cannot simply be labelled as a naïve and intransigent projectile theorist since, while remaining committed to the core of Newton's theory, both nature and his contemporaries forced him to deviate from Newton on some important issues. Secondly, it will be argued that he was not a dull empiricist but took a lively interest in theory, although he considered that some of his contemporaries had mistakenly over-emphasised theory. In proposing this more nuanced interpretation of Brewster I intend also to uncover his more basic presuppositions about science and suggest how these affected his response to innovations in optical theory, especially the wave theory and Thomas Young's principle of interference.

I

While a student at Edinburgh University in the 1790s, Brewster encountered the prevalent views about the projectile and wave theories of light[5].

Fig. 9. (a) and (b) Royal Medal, awarded to David Brewster by the Royal Society of London. (*A. G. L. Baxter, W.S.*).

However, it was not until 1806, when he was in his mid-20s, that he published his first paper on optics. There are nevertheless strong indications that his interest in physical optics was first kindled several years prior to this first publication. By 1798 he was grinding object glasses for telescopes[6]. Moreover, in her biography Mrs. Gordon tantalisingly informs us that in 1800 he made his first discovery: 'It was in his favourite science of Optics, discovering a new and important *fact*, while submitting the *Newtonian* [i.e. projectile] theory of light to a serious experimental test'[7]. Of this discovery and its significance Mrs. Gordon said nothing more. Another biographer was, however, more explicit in stating that in '1799 he was induced by his fellow student [Henry] Brougham, to study the inflection of light, repeating Newton's experiments'[8]. Other sources date an important discovery respecting inflection from the previous year, 1798[9]. Taken together these sparse references appear to indicate that Brewster was studying inflection in the late 1790s and made a discovery of some importance concerning theories of light.

The term inflection—what we could call diffraction—referred to the coloured fringes produced close to the edge of the shadow of an opaque body. Moreover, this term was theory-laden since on the projectile theory the rays passing close by a material body were conceived to be bent—inflected by a short-range force. Newton had suggested in the *Opticks* that inflection fringes were produced by a repulsive force[10]. Nearly a century later, in 1799, Gibbes Walker Jordan argued that Newton and others were mistaken and that the fringes were produced not by a repulsive but by an attractive force[11]. Some three years before Jordan's publication a precocious student at Edinburgh, Henry Brougham, had sent to the Royal Society of London a paper in which the inflective force was treated as similar in type to the refractive force but operating inversely on the spectral colours[12]. This was the disputed territory that Brewster entered, probably under Brougham's tutelage, in his earliest optical researches.

While he did not publish the results of this research and few documents remain from this period, a short manuscript dated 19 January 1802 exists which appears to summarise the fruits of his early labours and which sheds important light on his own intellectual development[13]. Attempting to gain patronage and the attention of the scientific community Brewster sent this paper to the Earl of Buchan whom he hoped would communicate it to the Royal Society of London. Buchan forwarded it to Sir William Herschel among whose papers it remains to this day, never having been published.

Brewster's 'Observations on the inflexion of light' is an unsophisticated production. While effectively criticising the accounts of inflection offered by other writers, Newton included, Brewster offered a few positive conclusions. For example, he suggested that a close connection existed between inflection and refraction, arguing that the deposition of colours was the same in both cases—a conclusion contrary to Brougham's. Also, in analysing the geometry of inflection he concluded that the trajectory of the inflected ray was such that the power of inflection was the 'simple inverse ratio of the distance', a conclusion which he derived from a dubious mathematical analysis. However, for the purpose of the present discussion, the most significant aspect of the paper was his criticism of some widely-accepted aspects of the projectile theorists' account of inflection. Not only did he follow Jordan in rejecting repulsive forces as responsible for inflection but he also undermined the often-assumed analogy between the optical force and the force of gravitation. In the gravitational case the force on a test body is proportioned to the mass of the body to which it is attracted. Hence a dense body would attract the test body more strongly than a rarer body of the same dimensions. However, in the case of inflection he discovered experimentally that the fringe pattern produced was 'universally the same' and did not depend on the density of the diffracting material. Thus by 1802 Brewster certainly had reason to doubt whether forces—as understood on the analogy with gravitation—were responsible for inflection. Brewster was in 1802 but a short step away from the position he made clear in 1817 that 'From the experiments on inflection, it follows that the deviation which the rays of light experience in passing by the edges of bodies, is not produced by any force inherent in the bodies themselves, but that it is a property of the light itself'[14]. Although he was not so explicit in 1802 it is clear not only that he was puzzled about the precise details of inflective forces but also that he was making little headway with analysing inflection in terms of short range forces. In this respect Brewster had already departed significantly from the standard eighteenth-century interpretation of projectile optics[15].

If, then, inflection fringes were not produced by short-range forces, how did Brewster account for the phenomena? His early researches seem to have prepared him for a very different solution to the problem. At the turn of the century Thomas Young had proposed a law of optical interference: two rays, he claimed, travelling from the same source along different paths, will interfere, producing brightness if the path difference is an integral number of wavelengths and darkness if the difference involves half wavelengths[16]. In a paper read before the Royal Society in 1801 Young had argued that inflection fringes were produced by the interference of two sets of rays, one travelling directly to the screen while the other is reflected from the inflecting body (which on his theory is surrounded by an atmosphere of denser ether). This explanation, he claimed, 'fully agrees with the observations of NEWTON'S third book, and with those of later writers'[17]. Although he subsequently changed his mind over some details of this explanation he frequently cited inflection fringes as a major confirmation of his principle of interference.

Considering Brewster's failure to account adequately for inflection by the projectile theory we would expect him to pay serious attention to Young's alternative explanation. This he did but only after a considerable time delay. Moreover, given that Brewster and Young were the two most impressive optical writers in Britain in the early nineteenth century we would also expect frequent references to Young in Brewster's numerous long papers published during the 1810s. Such reasonable expectations are not, however, realised. Even allowing for the limitations of the wave theory at that time, especially with respect to polarisation, there are many points throughout Brewster's first two dozen optical papers where references to Young's work would have been appropriate. For example, in a paper read before the Royal Society of Edinburgh in February 1815 Brewster refers to Newton's discovery of the phenomenon of thick plates and the subsequent research by the Duke de Chaulnes, Gibbes Walker Jordan and Henry Brougham[18]. Could Brewster have been ignorant of Young's contribution to that subject? It seems unlikely that he would not have read Young by 1813, but if he had it made little lasting impression on him. Only in the final paragraph of a paper read before the Royal Society in January 1818 did Brewster finally acknowledge Young's 'beautiful law of interference' which, he claimed, could explain not only the colours of polarised light but also an impressive range of phenomena including the colours produced by thin and thick plates and by inflection[19]. In later writings we also find many positive references to the principle of interference.

In attempting to account for Brewster's initial disinterest in Young's work and his later positive but partial advocacy of the principle of interference we need to trace his intellectual development to 1822 when his article 'Optics' was published in the *Edinburgh Encyclopaedia*. Most of Brewster's early papers are crowded with experimental results obtained while investigating the optical properties of one substance, then another. At times he seems intoxicated with the mere originality of his discoveries; however, the search for the more fundamental properties of matter was never far away. The word *law* appeared frequently in his papers; indeed he conceived the discovery of laws as the next analytical stage in scientific investigation. 'The establishment of a new law', he claimed, 'must at all times be considered as an important step in the progress of science, but when this law presides over a class of facts, all of which are unexplained, and many of which still remain to be discovered, it claims a higher regard both as an instrument of discovery, and as a principle for explaining new and analogous phenomena'[20]. His most famous law—Brewster's law, so called—connects refractive index with angle of polarisation, but it is only one of many[21]. In the early papers most of the laws cited concerned interrelations between light and different forms of matter. Moreover, for reasons to be discussed later, it is significant that in the 1818 paper already cited he referred to the principle of interference as a *law*.

Brewster's programme did not end with laws; rather laws provided a stepping stone towards the discovery of causes[22]. The discovery of the causes of optical phenomena, however, remained far in the future and he acknowledged that first many more facts and laws would have to be found. Moreover, his early papers exhibit a frequent if uneven predeliction for theoretical speculation.

'I have made use of no hypothetical assumptions', Brewster stated in 1815, but went on to claim that in 'imitation of MALUS, the language of theory has been occasionally employed, but the terms thus introduced are merely expressive of experimental results'[23]. Despite such assertions Brewster's language betrayed more of his theoretical presuppositions than he was prepared to acknowledge. Almost unselfconsciously the terminology of Newton's projectile theory began to enter: 'not one particle of it [light] will suffer reflection'; '*secondary poles* of extraordinary reflection'; 'attracting or refraction force ... repulsive or reflecting force'; 'the existence of such a [repulsive] force being unquestionable ...'—to cite but a few examples[24]. However, Brewster's concern with forces became far more self-conscious and strident towards the end of 1815. The context of this change was his research into double refraction and polarisation in which different forces were employed to account for the behaviour of the ordinary and extraordinary rays. His work on double refraction and polarisation needs to be interpreted against the background of research carried out in France by Malus and Biot, and it is likewise the French connection which is important in accounting for the development of his ideas about force. To Brewster's rather selfconscious attempt of early 1814 to account for the colours in mother of pearl[25]—an area in which Young's principle of interference might have been relevant—needs to be added his recognition by late 1815 that Biot and more importantly Laplace had elaborated force theories to account for double refraction[26]. At about the same time his conviction that polarising forces existed gained further strong support from his discovery that heat and mechanical stress could produce double refraction in glass. The application of external pressure could thus realign the forces internal to a piece of glass[27]. Thus force, if not forces, seemed responsible for light-matter interactions and he even frequently speculated on the close analogy between the optical, magnetic and electrical forces of polarisation. Thus although Brewster had encountered considerable problems in accounting for inflection by short-range forces, by 1816 he accepted unquestioningly the action of forces in explaining polarisation and double refraction.

The force theories of Newton and Laplace were not the only explanations of double refraction in Iceland spar; this phenomenon had also been the subject of Huygens' wave theory, a theory which Thomas Young had revitalised in several reviews published in the *Quarterly Review* between 1809 and 1814[28]. Throughout these reviews Young's principal contention was that Laplace's projectile account of double refraction was confused and based on gratuitous assumptions and that he had prematurely dismissed Huygens' theory. Young proceeded to show with the aid of some detailed examples that the Huygenian theory of ellipsoidal waves provides not only an adequate account of double refraction in Iceland spar but also that the theory could readily be extended to some of the other cases discussed by Malus.

Brewster first acknowledged Young in print in a note to a paper of May 1815 in which he cited one of Young's reviews as his source for Malus' recent work on double refraction[29]. A few months later Brewster cited an earlier review in which Young—'an eminent English philosopher'—had shown that an undulation will assume a spheroidal form in an isotropic stratified substance[30]. Likewise the published extracts from the Young–Brewster correspondence[31] also confirm the thesis that Brewster first took cognisance of Young through the reviews in the *Quarterly* and not his early papers in the *Philosophical Transactions* or his *Lectures* published in 1807. Moreover, the initial significance of Young's work for Brewster was as reviver and extender of the Huygenian theory of double refraction. Although Brewster never unequivocally adopted that theory, it appears that Young's reviews served to make him reconsider the theory and to acknowledge that it possessed some explanatory power.

While Young's reviews in the *Quarterly* were concerned principally with double refraction they do contain brief and very inadequate accounts of his principle of interference of light. Probably owing to his antipathy towards wave theories of light Brewster ignored the initial formulations of the principle (1802–7) and also the claims Young made for its explanatory power in the *Quarterly*. Instead he appears to have recurred to Newton's much criticised theory of fits; thus in a paper of February 1815 he expressed his conviction that a set of new fringes was 'explicable by the beautiful theory of Fits of easy reflection and transmission', although he did not actually compare theory with experiment[32]. In a letter a few months later he referred to Young's theory of 'recurrent colours'; a term Young had used in one of his reviews but hardly the term Brewster would have chosen had he been intimately familiar with the principle of interference[33]. Young's reply dated 13 September 1815 pressed Brewster hard to pay attention to the principle of interference, a principle which Young clearly believed Brewster to have overlooked. This letter is also significant because he explicitly dissociated interference from the wave theory. Young admitted that he had somewhat relinquished the wave theory *per se* but still he held steadfastly to his *laws* of interference 'since almost every new case of the production of colours, that has been lately discovered, ranges itself as a simple consequence of these laws, and is as regularly deducible from them by calculation, as the motions of the planets are deducible from the laws of gravitation'[34]. Although the distinction between laws and theories would have been familiar to Brewster, this interpretation of interference as a *law* (as distinguished from the wave *theory*) was made explicit to him in this correspondence. Given the importance which Brewster himself assigned to laws Young's argument should have appealed to him. Even so we have to wait over two years before the January 1818 paper in which, after further exchanges with Young, he paid tribute to the 'beautiful law of interference'.

When Brewster came to write his article 'Optics' (published 1822) for the *Edinburgh Encyclopaedia* he was full of praise for the law of interference which, he claimed, Young had founded firmly on experiment and which had been impressively confirmed by a number of recent discoveries[35]. Yet in the sections of the article in which Brewster discussed particular classes of phenomena references to optical interference were sparse. The sections on polarisation and double refraction were couched principally in terms of forces. More surprisingly, when discussing the colours of thin plates Brewster gave a full account of Newton's theory of fits merely noting in a short concluding paragraph that Newton's experimental results were also in agreement with deductions from the law of interference[36]. Significantly it was only in the section on inflection that Brewster gave considerable emphasis to the law of interference. Young's contribution, he admitted, had been substantial since he had discovered fringes *internal* to the shadow of a narrow piece of card produced by the interference of the light passing by both sides. 'Hence Dr. Young obtained an *experimental demonstration of the law of interference*: a law which though neglected at the time by many philosophers [Brewster included], and opposed by the ignorance and jealousy of others [Brougham?], has, like every other production of genius, triumphed over all opposition, and is now universally [*sic*] admitted as a general principle in physical optics'[37]. Despite this grand flourish it is clear that even in the early 1820s Brewster was prepared to acknowledge the law of interference primarily as a solution to the problems of inflection which he had encountered twenty years earlier. Evidence for this claim is to be found not only in the *Edinburgh Encyclopaedia* but also in a roughly contemporary report in the *Edinburgh Philosophical Journal* in which Brewster outlined Fresnel's recently-published paper on diffraction[38]. It was in the context of this report that Brewster admitted that his own work on inflection, dating from 1798 and 1812–13, had led him to reject the Newtonian theory of inflection and to the conclusion that 'inflexion was not produced by any force inherent in the bodies themselves'.

Even though Brewster's acceptance of the law of interference was both slow and partial he was still probably the first British natural philosopher even to pay it that much attention. Brewster's qualified acceptance of interference raises an historiographical point which deserves further clarification. Historians have traditionally conceptualised the development of physical optics as the confrontation between two opposing theories of light—the projectile and wave theories. However, among the many good reasons to abandon this simplistic dichotomy is the recognition that Brewster and Young could agree that the principle of interference was a *law* of nature totally independent of the ultimate nature of light. Thus while one adopted a wave and the other a projectile hypothesis both accepted that light behaved in a manner represented by the law of interference, a law expressible in simple mathematical terms. Brewster made the logical points that a true law could be deduced from any of a number of theories, and that the truth of the law does not imply the truth of the theory from which it is deduced. Thus, for example, the (true) law of double refraction in Iceland spar was deduced from the useful but false wave theory of Huygens. Again, the law of interference, although initially deduced from the wave theory, was theoretically independent. In his 1802 lecture notes Young had made the same point and had hinted at a mechanism for reconciling the projectile theory with the law of interference[39]. In the 1822 *Edinburgh Encyclopaedia* article, after admitting that the 'beautiful law of interference is not easily explained upon the Newtonian theory of emission' Brewster proceeded to elaborate a suggestion made by Young. The particles of light of a specific colour were supposed to follow each other at equal intervals while interference was explained in physiological terms. Thus, for example, destructive interference occurred when particles arrive at the eye at half-interval durations thus setting up two mutually destructive vibrations—180° out of phase with each other—in the optic nerve[40].

It is also significant that in his presentation of interference in 1822[41] Brewster did not utilise the term *wavelength*, which was inexorably tied to the wave theory, but instead his exposition was in terms of observables. Thus for a set of interference fringes the path difference, d, between two adjacent fringes could readily be ascertained. The law then implied that bright fringes would be located at path differences $2d$, $3d$, $4d$, ..., and dark bands at $\frac{d}{2}, \frac{3d}{2}, \frac{5d}{2} \ldots$.

Although Brewster tried hard to compare the wave and projectile theories of light in his 1822 article his own preference for the projectile theory was barely disguised. He had parted company from Newton and the traditional projectile theorists by rejecting a central role for short-range forces in accounting for inflection. While we should not underestimate the significance of this departure from tradition we should also recognise that his optical research was directed to discovering the interaction of light and different forms of matter. From the start of his research career these interactions were conceived in terms of particles and forces and his subsequent experience only increased his commitment to this paradigm. His important work on polarisation and double refraction was, as already discussed, largely conceived in terms of differential forces acting on the ordinary and extraordinary rays. Moreover, his commitment to the corpuscular nature of light drew considerable strength from chemistry since he firmly believed that it would be 'by the alliance ... of Chemistry with Optics, that great revolutions are yet to be effected in Physics'[42]. He conceived that each chemical substance exhibited its own crystalline structure which was dependent on the forces of affinity acting between the constituent particles. These forces were likewise responsible for the refraction, reflection and polarisation of incident light. Thus in principal, at least, the structure of substances could be ascertained from their chemical and optical properties. In Brewster's opinion this intimate connection between optical and chemical phenomena was not susceptible to explanation on the wave theory but strongly supported a corpuscular theory of light.

In his 1822 *Edinburgh Encyclopaedia* article Brewster wrote that the true 'nature of light is absolutely unknown, and probably will remain among the arcana of science'[43]. However it seems clear that while Brewster appreciated that the explanatory power of the wave theory was being increased, threatening 'to overwhelm the Newtonian doctrine of emission'[44], and potentially offering explanations of recondite phenomena of inflection, thin plates, double refraction and even polarisation, the core of the projectile theory which remained beyond challenge was the particulate nature of light. Even in the later controversies, which we now examine, he retained a personal conviction in the existence of light particles.

II

A few years after the publication of his *Edinburgh Encyclopaedia* article a wave theory of light, derived largely from Augustin Fresnel, not Young, and avidly supported by mathematically-trained men from Cambridge and Dublin, rose into prominence. In a flurry of technical literature the bounds of the new theory were rapidly extended as it was brought to account for an impressive range of intricate phenomena. The energetic advocates of the new theory mercilessly attacked projectile optics and some, too, urged the ultimate truth of their own doctrines. Disciples of the new creed also had a social advantage since George Airy, Humphrey Lloyd and Baden Powell filled natural philosophy chairs at the major universities while others—like John Herschel and William Whewell—were effective publicists. By contrast Brewster did not hold a teaching post and thus could not train students in projectile optics: he did, however, wield the reviewer's pen against his opponents. Moreover, he had a few allies; principally Richard Barton (*d*. 1834), known to posterity for patenting 'Barton's buttons', Richard Potter, the incompetent professor of natural philosophy at University College, London, and, later, Lord Henry Brougham, the ageing ex-Lord Chancellor. In the ensuing dispute[45], which only died with the protagonists, we can identify four related issues—namely, substantive, methodological, stylistic and sociological concerns.

On the first of these issues—the substantive—Brewster made his position abundantly clear. The new wave theory was superior to the projectile theory since the former could explain a wide range of phenomena with breath-taking accuracy while the explanatory power of the latter was fairly limited. However, despite its impressive successes the wave theory was inadequate in several areas, selective absorption and refractive dispersion being the two best-known. Moreover, the wave theory did not account for photo-chemical phenomena such as the blackening of silver salts. (Since the wave theorists did not consider that their theory should have explained the last class of phenomena we have here an example of the protagonists differently defining the domain of optical theory.) Thus while ready to acknowledge the strengths of the wave theory he refused to overlook its limitations. One can detect a hardening of Brewster's attitude as he became increasingly isolated and threatened during the 1830s and 1840s; the more his opponents sang the praises of the wave theory the more he pressed those phenomena it did not explain—both those outside its scope and those it incorrectly predicted.

The failure of the wave theory to explain certain phenomena brings us to our second issue—methodology. As E. W. Morse has argued, Brewster considered that science enabled man to discover how God had constructed the universe[46]. In implementing this notion of science he sharply distinguished true causes from baseless speculations. A necessary condition for a true cause was that it had to explain *all* the phenomena within its domain; failure to do so indicated that the assigned cause was false. Thus many of Brewster's attacks on the wave theory were aimed at showing that its deductions were falsified in experimental situations. Hence in a number of his papers he delineated 'new' phenomena which apparently falsified the theory[47]. A related strategy, which he principally employed in his review articles, was to identify classes of phenomena—such as refractive dispersion and selective absorption—for which the wave theory offered no account[48]. These various arguments provided Brewster with strong support for his claim that it had not been proved that light is a wave motion.

Brewster's methodological presuppositions provided him with further weapons to attack his opponents. As has already been noted he keenly advocated Young's *law* of interference. This law was on a par with the law of gravitation since both were descriptions of how nature always behaved. Brewster, like many other Scots of the late eighteenth and early nineteenth centuries, contrasted such laws with physical causes involving some hypothesis about the actual structure of nature. Brewster had no objection to the employment of hypotheses. His own research contained numerous speculations and he recognised that hypotheses, such as Huygens' and Young's wave theories, had been very useful in the discovery of new optical laws. Thus when Auguste Comte proposed the elimination of all causal hypotheses Brewster sprang to their defence. Not only, he argued, had they proved very helpful in aiding the memory and in explaining facts but, more importantly, some hypotheses had positively aided the advancement of science by indicating new laws and phenomena:

'Now, though the undulatory theory does assume an *ether*, invisible, intangible, imponderable, inseparable from all bodies, and extending from our own eye to the remotest verge of the starry heavens; yet as the expounder of phenomena the most complex, and otherwise inexplicable; and as the predicter of highly important facts, it must contain among its assumptions (though, as a physical theory, it may still be false) some principle which is inherent in, and inseparable from, the real producing cause of the phenomena of light; and to this extent it is worthy of our adoption as a valuable instrument of discovery, and of our admiration as an ingenious and fertile philosophical conception'[49].

As the above quotation makes abundantly clear, Brewster was impressed by the power of the wave theory while unwilling to accept it as the true theory of light. When his opponents sang its praise and, in his opinion, grossly overstated their case he sought to undermine their claims. In several reviews he repeated his theme: the wave theory is only an hypothesis but one which has been dogmatically supported by its proponents[50].

Brewster considered that unlike the law of interference the wave theory of light was merely a physical hypothesis since it failed to fulfil the two necessary conditions laid down by the *vera causa* principle—sufficiency and truth. Firstly, as already pointed out, it was not *sufficient* to explain the phenomena. Secondly, it involved an hypothetical luminiferous ether which was 'invisible, intangible, imponderable, [and] inseparable from all bodies'. Hence Brewster followed Reid, Brougham and a number of other Scots in rejecting ether theories as the true causes of natural phenomena[51]. This attack on ether theories figures prominently in his reviews and he particularly objected to those like Whewell who, abandoning all philosophical caution, asserted that ethers really existed. Yet Brewster's antipathy towards ethers was not merely owing to their unobservability and thus their purely hypothetical status but also to the implication that in hypothesising the structure of nature man committed the sin of pride and forsook the study of God's works. As he wrote in one of his reviews, 'we shall not forget that it is a hypothesis we have selected; and we shall never endeavour to magnify the divine wisdom, by inviting man to admire his own hypothesis, and by loading our Natural Theology with the lumber of human wisdom'[52].

The above discussion raises the question whether Brewster was consistent in applying the same methodological criteria to both theories. While it is difficult to give a definitive answer to this question (since Brewster not only retained aspects of the language of projectile optics but also continued to speculate on the nature of light) it is significant that he refused to commit himself in public to either theory, both being unable to explain all optical phenomena. Moreover, despite being well disposed towards the projectile theory Brewster later followed Biot and Brougham in describing himself as a 'rieniste'[53].

Consideration of these methodological issues leads us to the related problem of style. Given his aversion to the overemphasis on hypotheses Brewster was strongly opposed to the attitude of those—like the Cambridge-trained mathematical physicists—who constructed detailed mathematical models which, only after long trains of deductive reasoning, encountered empirical reality. Mathematics, for Brewster, certainly provided a means of expressing empirical laws but the language of higher mathematics not only proved difficult for him to understand but also enshrined a false hypothesis-based methodology. By contrast Brewster asserted the importance of experiment from which inductive generalisations, often expressible in mathematical terms, should be sought. He greatly valued experiments in themselves since he believed they revealed truths of God's creation. His deep commitment to the discovery of experimental truths involved a great personal sacrifice; not only did he spend much of his life in the laboratory and much money on apparatus, but he was blinded for several weeks after a chemical experiment exploded and also suffered a number of other accidents[54]. He was, in his own opinion, a true martyr of science and could not comprehend those who failed to value his experimental work. When the Royal Society refused to publish one of his papers his first line of defence was to claim that his article was 'full of new [experimental] matter'[55]. Again, his advice to Brougham on a paper the latter had communicated to the Royal Society was that since it 'contains positive Experiments and Observations, I do not see how it is possible that they can refuse to place it in the Transactions'[56]. Moreover, one of his complaints against the wave theorists was that they were mere theoreticians who were unfamiliar with the optical phenomena they discussed and with the complex experimental techniques involved[57]. By contrast, in Brewster's Calvinistic ethic the experimenter was the honest artisan of science who discovered new and true facts.

By the early 1840s Brewster had become an embittered, lonely outcast from British optics. The battle lines dividing him from his opponents were not merely concerned with the wave theory of light but spread to many other areas. He was a Scot, his opponents principally English: there was a generation gap between them: he was an experimentalist, they were mathematically-trained theoreticians: he failed to gain a chair, whereas many of his opponents taught in universities and attracted disciples: he was an Evangelical, they were mostly broad churchmen[58]. The list can, of course, be extended. While we have already briefly touched on some of these issues we shall now concentrate on the widening social and institutional divide separating Brewster from other optical writers.

The history of the 1802 manuscript on inflection set a precedent for Brewster's later dealings with London, if not English, science. His first contact with William Herschel could have done little to enhance the young man's self esteem. Herschel eventually sent back '50 annotations', most of which, one imagines, contained destructive criticisms. Moreover, Herschel's attitude was authoritarian, over-bearing and hardly encouraging[59]. He did not lay Brewster's paper before the Royal Society and it was a decade before Brewster again approached the Society.

There followed a period of cordial relationships between Brewster and London-based scientists. However, in the early 1840s, Brewster found the door of the Royal Society slammed shut in his face. Writing to Brougham he claimed that the wave theory had become 'the Creed of the Society'[60], and in letters to the Treasurer and two Secretaries he complained that the anonymous referee who had rejected one of his papers was Airy who 'has acted from personal feelings, he & Whewell having privately and publicly done all in their power to depreciate my labours'. Moreover, he complained that the 'Undulatory Theory is now to be the test of all Optical Papers submitted to the Royal Society. If the most important discoveries are made, they are considered of value only if the Undulationists can explain them. If not they are worthless'[61].

Rebuffed by the 'Cambridge faction hostile to all Scotsmen'[62], Brewster almost entirely severed his connections with the Royal Society of London, the rejected paper being later published by the Royal Irish Academy[63]. Brewster subsequently presented much of his optical research to the Royal Society of Edinburgh and the British Association, the latter being a broader public forum before which he tried to discredit his opponents, as described in J. B. Morrell's article in this volume. The *Philosophical Magazine* and review periodicals also furnished suitable media for extending the dispute.

Commencing in the early 1840s Brewster encouraged his old colleague Brougham to 'take up the subject of the Emission *versus* the *Undulatory* theory of light. In this cause I have struggled single handed for the last 30 years'[64]. The correspondence between these two ageing natural philosophers is revealing, if at times pathetic. Brougham, who forty years earlier had rejected Young's law of interference, was tutored by Brewster in the recent developments in optical science and assured that the law of interference had to be accepted[65]. After carrying out a number of inflection experiments Brougham, urged on by Brewster, submitted three papers to the Royal Society of London. The first was published in the *Philosophical Transactions* for 1850 despite protests from Airy[66]. However Airy, together with George Stokes, refused to accept the second paper two years later and its publication was finally rejected by John Herschel[67]. Airy's report on the paper reflects the assessment by the new regime of both Brougham and Brewster. He dismissed the paper as containing trivial experimental results which, he considered, were perfectly conformable to the wave theory. Moreover, 'it is written in entire ignorance both of the general principles of the great Undulatory Theory and of the algebraical and numerical results which have been deduced from it. The paper is more than twenty years behind the actual state of science'[68].

III

If Brewster ended his career in physical optics embittered and rejected by the new regime we should perhaps recognise him as a casualty of a 'scientific revolution'. From this perspective he emerges as a representative of an earlier theory of light and an outmoded style of science. Thus one historian draws the contrast between Fresnel, who 'attacked the existing theory head on and created a revolution in optics', and Brewster, who 'conformed his work to the existing system' and 'confined himself to filling in the minutiae'[69]. However, this contrast between Fresnel and Brewster is too sharply drawn so that Brewster is portrayed as a dull experimentalist unconcerned with theoretical innovation[70] or with philosophical niceties. While there were clearly immense and important differences between these two physicists this prevailing account of Brewster needs to be modified considerably, in at least four respects.

Firstly, it has been shown that despite his excessive reverence for Newton Brewster departed significantly from the general interpretation of projectile optics by rejecting the role of short-range forces in accounting for inflection. *Secondly*, it is important to recognise that Brewster was probably alone in the first two decades of the nineteenth century in recognising the power of Young's law of interference, even though his adoption of the law was both slow and partial. *Thirdly*, Brewster was not opposed to theory but adopted an intelligent and sophisticated attitude to the relation between experiments, laws and theories. Moreover, his optical writings indicate his deep interest in the opposing theories of light. *Finally*, far from being simply opposed to the wave theory or ignoring it, Brewster both appreciated its explanatory power while remaining critical of some of the claims made by its proponents.

Bicentenaries are appropriate occasions for re-evaluating a writer's work, and I have taken the opportunity offered by Brewster's bicentenary to build on the work of E. W. Morse and others[71] by re-evaluating Brewster as a thinker and by showing that in at least the four ways discussed above he deserves a not insignificant chapter in the history of theories of light.

Notes and References
For permission to quote from manuscripts in their possession the author gratefully acknowledges the Royal Society of London, the Royal Astronomical Society and University College, London.

1. Anon., 'Brewster, Sir David', *Encyclopaedia Britannica*, 24 vols. (9th edition, Edinburgh, 1875–1889), IV, 276.
2. On optical theories in the late eighteenth and early nineteenth centuries see W. Whewell, *History of the Inductive Sciences, from the Earliest to the Present Times*, 3 vols. (London, 1837) II, 330–462; H. J. Steffens, *The Development of Newtonian Optics in England* (New York, 1977) which contains a useful chapter on Brewster; G. N. Cantor, 'The Reception of the Wave Theory of Light in Britain: A Case Study Illustrating the Role of Methodology in Scientific Debate', *Historical Studies in the Physical Sciences*, 6 (1975), 109–132; G. N. Cantor, *Optics after Newton: Theories of Light in Britain and Ireland, 1704–1840* (Manchester, 1983).
3. M. M. Gordon, *The Home Life of Sir David Brewster* (2nd edition, Edinburgh, 1870), 263; R. S. Westfall, 'Introduction' to D. Brewster, *Memories of the Life, Writings and Discoveries of Sir Isaac Newton*, 2 vols. (New York and London, 1965), I, xxvi–xxix.
4. Ibid., xxviii–xxix.
5. Brewster's name appears on the matriculation list for John Robison's natural philosophy class in 1798. Judging from his lecture notes (Edinburgh University Library, Ms. Dc. 7.24) Robison extensively discussed both theories of light, although he strongly supported the projectile theory against the wave theory.
6. Correspondence with James Veitch, see handlist in National Register of Archives (Scotland), Survey 337.
7. Gordon, *op. cit.* (3), 46; emphases added. *Cf.* Westfall, *op. cit.* (3), xiii.
8. Anon., *op. cit.* (1), 276.
9. See report of the proceedings of the Royal Society of Edinburgh in *The Quarterly Journal of Literature, Science and the Arts*, 2 (1817), 207–208; D. Brewster, 'Account of M. Fresnel's Discoveries Respecting the Inflexion of Light', *Edinburgh Philosophical Journal*, 2 (1820), 150–153. See also E. W. Morse's informative and well-researched account 'David Brewster' in C. C. Gillispie (ed.), *Dictionary of Scientific Biography*, 16 vols. (New York, 1970–76), II, 451–454.
10. I. Newton, *Opticks* (London, 1704), book III. See also query 21 (subsequently query 29), added to the first Latin (1706) edition.
11. G. W. Jordan, *The Observations of Newton concerning the Inflections of Light; Accompanied by other Observations differing from his; and appearing to lead to a change of his Theory of Light and Colours* (London, 1799).
12. H. Brougham, 'Experiments and Observations on the Inflection, Reflection, and Colours of Light', *Philosophical Transactions of the Royal Society of London*, 86 (1796), 227–277. See also his 'Further Experiments and Observations on the Affections and Properties of Light', *ibid.*, 87 (1797), 352–385.
13. The manuscript, entitled 'Observations on the Inflexion of Light', is prefaced by a letter from Brewster to the Earl of Buchan. Royal Astronomical Society, Ms Herschel, W., 6/17.
14. 'Proceedings ...', *op. cit.* (9), 207.
15. Cantor, *Optics ... op. cit.* (2), 25–90.
16. Slightly different accounts of the principle of interference appeared in T. Young, *A Syllabus of a Course of Lectures on Natural and Experimental Philosophy* (London, 1802); T. Young, 'On the Theory of Light and Colours', *Philosophical Transactions of the Royal Society of London*, 92 (1802), 12–48; T. Young, 'An Account of some Cases of the Production of Colours not hitherto Observed', *ibid.*, 92 (1802), 387–397; T. Young, 'Experiments and Calculations Relative to Physical Optics', *ibid.*, 94 (1804), 1–16; T. Young, *A Course of Lectures*

on Natural Philosophy and the Mechanical Arts, 2 vols. (London, 1807), I, 457–471.
17. Young, 'On the theory . . .', *op. cit.* (16), 43.
18. D. Brewster, 'On a New Species of Coloured Fringes Produced by the Reflexion of Light between Two Plates of Parallel Glass of Equal Thickness', *Transactions of the Royal Society of Edinburgh*, 7 (1815), 435–444.
19. D. Brewster, 'On the Laws of Polarisation and Double Refraction in Regularly Crystallized Bodies', *Philosophical Transactions of the Royal Society of London*, 108 (1818), 199–273, esp. 270–273.
20. D. Brewster, 'On the Polarisation of Light by Oblique Transmission through all Bodies, whether Crystallized or Uncrystallized', *Philosophical Transactions of the Royal Society of London*, 104 (1814), 219–230, esp. 229.
21. *Ibid.*, 221; D. Brewster, 'On New Properties of Light Exhibited in the Optical Phenomena of Mother of Pearl, and other Bodies to which the Superficial Structure of that Substance can be Communicated', *Philosophical Transactions of the Royal Society of London*, 104 (1814), 397–418, esp. 401; D. Brewster, 'On the Laws which Regulate the Polarisation of Light by Reflexion from Transparent Bodies', *ibid.*, 105 (1815), 125–159, esp. 125; D. Brewster, 'On the Laws which Regulate the Distribution of the Polarising Force in Plates, Tubes, and Cylinders of Glass, that have Received the Polarising Structure', *Transactions of the Royal Society of Edinburgh*, 8 (1818), 353–372; D. Brewster, 'On the Laws of Polarisation and Double Refraction in Regularly Crystallized Bodies', *Philosophical Transactions of the Royal Society of London*, 108 (1818), 199–273, esp. 205; D. Brewster, 'On the Laws which Regulate the Absorption of Polarised Light in Doubly Refracting Crystals', *ibid.*, 109 (1819), 11–28.
22. Brewster, 'On the . . . Polarisation of Light by Reflexion . . .', *op. cit.* (21), 158.
23. *Ibid.*
24. Brewster, *op. cit.* (20), 227; Brewster, 'On New Properties . . .' *op. cit.* (21), 404, 411 and 412.
25. *Ibid.*
26. D. Brewster, 'On the Optical Properties of Muriate of Soda, Fluate of Lime, and the Diamond, as Exhibited in their Action upon Polarised Light', *Transactions of the Royal Society of Edinburgh*, 8 (1818), 157–164. Brewster was, however, critical of some details of Biot's and Laplace's theories.
27. Brewster, 'On the . . . Distribution of the Polarising Force . . .', *op. cit.* (21).
28. [T. Young], review of Laplace's 'Sur la Loi de la Refraction Extraordinaire dans les Cristaux Diaphanes', *Quarterly Review*, 2 (1809), 337–348; [T. Young], review of 'Mémoires de Physique et de Chimie, de la Société d'Arcueil', *Quarterly Review*, 3 (1810), 462–481; [T. Young], review of works by Malus, Biot, Seebeck and Brewster, *Quarterly Review*, 11 (1814), 42–56.
29. D. Brewster, 'On the Multiplication of Images and the Colours which Accompany them in Some Specimens of Calcareous Spar', *Philosophical Transactions of the Royal Society of London*, 105 (1815), 270–293.
30. D. Brewster, 'On the Communication of the Structure of Doubly Refracting Crystals to Glass, Muriate of Soda, Fluor Spar, and other Substances, by Mechanical Compression and Dilation', *Philosophical Transactions of the Royal Society of London*, 106 (1816), 156–178, esp. 178.
31. G. Peacock and J. Leitch (eds.), *Miscellaneous Works of Thomas Young*, 3 vols. (London, 1855), II, 359–373.
32. Brewster, *op. cit.* (18).
33. D. Brewster to T. Young, 28 July 1815: Peacock and Leitch, *op. cit.* (31), II, 359–60. See also Young's review of Malus, etc., *op. cit.* (28), 49.
34. T. Young to D. Brewster, 13 September, 1815: Peacock and Leitch, *op. cit.* (31), II, 360–364.
35. [D. Brewster], 'Optics' [1822], for *Edinburgh Encyclopaedia* 20 vols. (Edinburgh, 1830), XV, 460–662, esp. 473 and 484–486.
36. *Ibid.*, 565.
37. *Ibid.*, 554.
38. Brewster, *op. cit.* (9).
39. Young's lecture notebooks: University College, London, Ms. Add. 13, 16/15r.
40. Brewster, *op. cit.* (35), 554.
41. *Ibid.*
42. D. Brewster, 'On the Optical Properties of Sulphurate of Carbon, Carbonate of Barytes, and Nitrate of Potash, with Inferences Respecting the Structure of Doubly Refracting Crystals', *Transactions of the Royal Society of Edinburgh*, 7 (1815), 285–302.
43. Brewster, *op. cit.* (35), 499.
44. *Ibid.*, 473.
45. Cantor, 'Reception of the Wave Theory . . .', *op. cit.* (2); T. L. Hankins, *Sir William Rowan Hamilton* (Baltimore & London, 1980), 131–154; J. B. Morrell and A. Thackray, *Gentlemen of Science: Early Years of the British Association for the Advancement of Science* (Oxford, 1981), 466–472.
46. E. W. Morse, 'Natural philosophy, hypotheses and impiety: Sir David Brewster confronts the undulatory theory of light', unpublished Ph.D. thesis, (University of California, Berkeley, 1972).
47. For example, D. Brewster, 'Report on the Recent Progress of Optics' *Report of the British Association for the Advancement of Science . . . 1832* (London, 1833), 308–322; D. Brewster, 'On a New Property of Light', *ibid.*, 7 (1837), 12–13; D. Brewster, 'On a New Property of the Rays of the Spectrum, with Observations on the Explanation of it given by the Astronomer Royal, on the Principles of the Undulatory Theory', *ibid.*, 12 (1842), 12.
48. [D. Brewster], review of W. Whewell's 'History of the Inductive Sciences', *Edinburgh Review*, 66 (1837), 110–151; [D. Brewster], review of W. Whewell's 'Philosophy of the Inductive Sciences', *Edinburgh Review*, 74 (1843), 265–306.
49. [D. Brewster], review of A. Comte's 'Cours de Philosophie Positive', *Edinburgh Review*, 67 (1838), 271–308, esp. 306.
50. *Ibid.*, 307–308; reviews cited in note 48; [D. Brewster], review of W. Whewell's 'Astronomy and General Physics Considered with Reference to Natural Theology', *Edinburgh Review*, 58 (1834), 427–457; [D. Brewster], review of M. Somerville's 'On the Connexion of the Physical Sciences', *Edinburgh Review*, 59 (1834), 154–171.
51. R. Olson, *Scottish Philosophy and British Physics, 1780–1880* (Princeton, N. J., 1975); L. L. Laudan, 'The Medium and its Message: a Study of Some Philosophical Controversies about Ether', in G. N. Cantor and M. J. S. Hodge, (eds.) *Conceptions of Ether: Studies in the History of Ether Theories 1740–1900* (Cambridge, 1981), 157–186; G. N. Cantor 'Henry Brougham and the Scottish Methodological Tradition', *Studies in History and Philosophy of Science*, 2 (1971), 69–89.
52. Brewster, Whewell's 'Astronomy', *op. cit.* (50), 456; G. N. Cantor, 'The Theological Significance of Ethers' in Cantor and Hodge, *op. cit.* (51) 135–155. See also Morse *op. cit.* (46) and John Tyndall, *Six Lectures on Light* (4th edition, London, 1885), 48, where Tyndall recounts that Brewster 'said to me that his chief objection to the undulatory theory of light was that he could not think the Creator guilty of so clumsy a contrivance as the filling of space with ether in order to produce light'. Quoted also by Steffens, *op. cit.* (2), 147.
53. [D. Brewster], review of three works by Arago, *North British Review*, 20 (1854), 459–500, esp. 488.
54. Gordon, *op. cit.* (3), 154, 170, 236.
55. D. Brewster to J. W. Lubbock, 25 June 1841: Royal Society of London, Ms. LUBB. 429.
56. D. Brewster to H. Brougham, 15 February 1850: University College, London, Brougham Archive, Ms. 26,656.
57. Brewster, *op. cit.* (53), 488.
58. J. H. Brooke, 'Natural Theology and the Plurality of Worlds: Observations on the Brewster-Whewell Debate', *Annals of Science*, 34 (1977), 221–286; S. F. Cannon, *Science in Culture: The Early Victorian Period* (New York & Folkestone, 1978).
59. W. Herschel to D. Brewster, March 1803: Royal Astronomical Society, Ms. Herschel, W., 1, 247.
60. D. Brewster to H. Brougham, 14 December 1841: University College, London, Brougham Archive, Ms. 26,624.
61. D. Brewster to P. M. Roget, 9 October 1841: Royal Society of London, Ms. M.l.3.189.
62. Brewster to Brougham, *loc. cit.* (60).

63. D. Brewster, 'On the Compensations of Polarised Light, with a Description of a Polarimeter for Measuring Degrees of Polarisation', *Transactions of the Royal Irish Academy*, 19 (1843), 377–392.
64. D. Brewster to H. Brougham, 26 September 1847: University College, London, Brougham Archive, Ms. 26,632.
65. See Brougham Archive, University College, London, Mss. 26,632–26,643 covering the period September 1847 to August 1849. A number of these letters concern inflection; for example in Ms. 26,634 (29 August 1848) Brewster urges Brougham to accept the law of interference as 'an experimental truth which the doctrine of undulations happens to explain'.
66. G. B. Airy on H. Brougham's 'Further Experiments on Light', 17 June 1852: Royal Society of London, Ms. RR. 2.36.
67. J. F. W. Herschel on H. Brougham's 'Further Experiments on Light', 1 November 1852: Royal Society of London, Ms. RR. 2.37.
68. Airy, *loc. cit.* (66).
69. Westfall, *op. cit.* (3), xxvi–xxix.
70. Brewster 'was not involved with the wave theory one way or the other': *ibid.*, xxvii.
71. Morse, *op. cit.* (9 & 46); Olson, *op cit.* (51); Steffens, *op. cit.* (2); Cantor, *op. cit.* (2).

Sir David Brewster: some concluding remarks

NICHOLAS PHILLIPSON

The papers that were given at this Symposium are an important preliminary survey of the territory that historians will have to cultivate if they are to make sense of the life and works of Sir David Brewster. What will have struck every reader of these papers—as it struck those who attended the Symposium—is the range of Brewster's activities in science and public life. Not only was he a student of physics and optics; he was a scientific instrument-maker, a prolific and successful scientific man of letters, admired for his editing of scientific journals, for his biographies of great scientists and, above all, for his popular scientific journalism. It is also apparent that he was an energetic and apparently successful operator in the highly politicised worlds of institutionalised science and university government. Overall, he has come over as a devout, domineering, competitive man of great intelligence, learning and energy, much admired by contemporaries as a grandee of the world of learning and as a scientist whose distinction was unmarked by the innovative qualities of greatness we associate with a Clerk Maxwell or a Kelvin. He was, in other words, a man of his time, not a man for all seasons.

Brewster is clearly a fascinating subject for a student of the early Victorian mind and it is a tragedy that the bulk of his papers were destroyed by fire earlier this century. A major theme of his story is upward mobility. For Brewster was a man of relatively humble origins (his father was rector of Jedburgh Academy) who spent his early years in the relatively confined circumstances of a small Scottish town. He sought fame and fortune in the classic Scottish manner, by exploiting the educational opportunities offered by the Scottish educational system and the highly competitive world of Edinburgh letters in the early decades of the century. For although there was not much money to be made here what was on offer was the prestige and sheer excitement of one of the major centres of intellectual activity in the western world and that in itself could serve as a springboard to future advancement in public life, in the professions, in politics or in government.

It is worth remembering what was required of the ambitious Edinburgh intellectual in the age of the late Enlightenment. Learning for learning's sake was not in the least valued. Indeed it might well be taken to be a mark of a socially irresponsible aestheticism or even of the sort of eccentricity of mind which was likely to warp intellectual judgement and betray a lack of investigative rigour. For learning was valued only inasmuch as it contributed directly to an enrichment of the mind of an inquiring citizen who was anxious to understand his duties as a man, citizen and, perhaps, as a servant of God. Thus a serious and ambitious intellectual was expected to be alert to the moral and theological implications of his discoveries and to their implications for rooting out moral, political and theological as well as scientific error.

It is of great importance to realise that these were not in any way secondary responsibilities, tiresome proprieties to be observed by men who could not afford to ignore a local audience; they were not, so to speak mere moral cherries on the cake of pure learning. These imperatives had long moulded the *mentalités* of Scottish intellectuals, shaping the way in which they selected problems for investigation and decided how to handle them. And to ignore them is to distort the intellectual history of the Scottish Enlightenment. This Symposium has given us ample evidence of Brewster's sensitivity to such imperatives and it raises the most important and interesting questions about the ways in which they shaped his scientific mind. We need to know much more than we do about the ways in which he chose and handled problems. We need to know whether, and if so what, tension there was between the demands of scientific investigation and

Fig. 10. Calton Hill, Edinburgh, *c*.1840. From a daguerreotype by an unknown photographer. (*Royal Scottish Museum*).

those of practical morality and theology. And we need to know whether such a tension generated innovation, conservatism or even eccentricity in his handling of purely technical problems. It would clearly be a gross error to underplay the depth of Brewster's piety or to try to isolate it from his science. Here, the evidence of his daughter and biographer, Mrs. Gordon is of great interest. She was a woman of intelligence and perception who had a gift for observing religious behaviour closely and exactly—a skill which our age has lost altogether. When she writes of her father engrossed in the Bible or at prayer or racked by religious anguish alone at night in his laboratory, she described the behaviour of a man who believed that there could not, *must* not, be any antagonism between religion and science. She recalls him asking 'How can we love Him, one whom we have never seen? We admire Him in His works and trust from the wisdom seen in these that He is wise in all His dealings—but how can we LOVE Him?' She replied that she had learned to love an invisible God. He took her in his arms, kissed her and replied 'in such a child like manner ... Go now then, and pray that I may know it too'[1]. In this story we hear the agonised voice of the Evangelical Christian who believed that while it was possible to prove the existence of God from the evidences of science, it was not possible to prove his love and beneficence.

It seems reasonably clear that Brewster's scientific thinking was shaped at every level by his life-long preoccupation with natural religion—that is to say, by the conviction that proofs of the being and nature of God could be adduced from the evidence of the natural world. For him the heart of the problem lay in understanding the structure of organised matter and, more particularly with light. At every point in his career he was driven by a conviction that the ascent from the world of experience to that of the infinite must be conducted by means of ever more minute, meticulous observation of particles which were imperceptible to the naked eye. Only such observation was good enough for science and religion. 'Speculation engenders doubts', he wrote. 'And doubt is frequently the parent either of apathy or impiety'[2]. But careful observation was impossible unless the eye and its workings were properly understood. As he wrote in *Good Words*, a popular and influential Evangelical review,

> 'Although every part of the human frame has been fashioned by the same Divine hand and exhibits the most marvellous and beneficent adaptions for the use of men, the human eye stands pre-eminent above them all as the light of the body and the organ by which we become acquainted with the minutest and the nearest, the largest and most remote of the Creator's work'[3].

One or two papers in this volume give the impression that Brewster merely set out to 'confirm' the truths of natural religion. But this is misleading for it implies that these truths were already known to him. Brewster was a man who believed that the resources of modern science now allowed men to penetrate the structure of the natural world in ways which were quite unprecedented and allowed men to reach an understanding of God's nature and purposes and of men's duties towards Him that had simply not been available to earlier ages. It is this powerful, quasi-mystical urge to explore higher and more absolute truths which seems to direct Brewster's science at all levels, from his studies of light, to his instrument-making and his popular scientific journalism. And it is something of which a future biographer will have to take careful stock.

All of this suggests a wider context for the study of Brewster's career, the history of Scottish Evangelicalism and the Disruption of 1843. Brewster was not raised in the simple country zealotry of rural Scotland. As a student at Edinburgh and as a young hopeful in the literary world, he joined the circle of Thomas Chalmers who had set out to lay the foundation of a new, polite, urban evangelicalism which was designed to take stock of the findings of modern philosophy and science, not to fly in its face. Brewster and the philosopher James McCosh, the future President of Princeton University, were to become the two most formidable intellectuals in Chalmers' circle in the 1830s[4]. Between them, they managed to develop a powerful Evangelical critique of the mental, moral and natural philosophy of the Scottish Enlightenment. In their view, the Scottish Enlightenment had taught men to judge reality from the simple appearance of things. No doubt they had been able to show that such a world was indeed ordered in its way but it was an order which presupposed that there was little meaning or purpose in existence. It was a lunar landscape which gave men little hope that they could find happiness in this world or salvation in the next. Brewster and McCosh set out to travel beyond the philosophy of the Enlightenment in order to map out an Evangelical landscape which gave evidence of beneficence as well as order, hope for the future as well as encouragement for the present. By so doing, they would give Evangelical Christianity an intellectual strength and a moral vigour which would release it from the hands of its two great enemies, country zealotry and the barren wastes of indifference most forcibly represented by the moderatism of established Scottish Presbyterianism.

But their efforts failed. Chalmers died before he or his disciples could take control of the new Free Kirk, created in the wake of the Disruption of 1843. Within a decade, control had passed to the old-fashioned Evangelicals and to neo-scholastic professors of theology who had little time for a faith rooted in the increasingly treacherous marshes of natural theology. McCosh went to America and to Princeton to turn that strict Presbyterian college into a powerhouse of Chalmersian Evangelicalism and into one of the great universities of *post-bellum* America. Brewster had less good fortune. At St. Andrews, his career as Principal was paralysed by bitter feuds for which he appears to have been considerably responsible. By the time he reached Edinburgh, he was an older, more withdrawn man, content to rest on his considerable laurels as a Scottish intellectual grandee. And his daughter's biography suggests that he was relatively content to do so.

Notes and References
1. M. M. Gordon, *The Home Life of Sir David Brewster* (Edinburgh, 1869), 315–317.
2. Quoted by E. W. Morse, 'David Brewster' in C. C. Gillispie (ed.), *Dictionary of Scientific Biography* (New York, 1970–76), II, 451–454.
3. D. Brewster, 'The Eye—Its Structure and Powers', *Good Words*, 3 (1862), 170–176.
4. I have dealt at greater length with McCosh in 'The Evangelist of Common Sense', *The Times Literary Supplement*, 23 October 1981.

Appendix I (a)

Scientific Apparatus associated with Sir David Brewster: An Illustrated Catalogue of the Bicentenary Display at the Royal Scottish Museum 21 November 1981–9 April 1982

A. D. MORRISON-LOW

This exhibition, brought together for the Symposium itself and running for a further four months, illustrated the main strands of Brewster's scientific work, namely his inventions, his research interests and his influences upon others[1]. Such an exhibition was, however, hampered by a lack of material closely related to Brewster, since his personal apparatus was destroyed with the Brewster family home in 1903. Items with known Brewster provenance are few, but from what remains it can be concluded that his own apparatus was not elaborate: his financial situation was straitened and he frequently tried to organise bodies such as the Royal Society of London[2], the Royal Society of Edinburgh[3], the University of St. Andrews[4], and the British Association for the Advancement of Science[5] to buy apparatus for his own use. His daughter's description of her father's study does not exaggerate its poverty:

> 'Much of his apparatus to unlearned eyes appeared a mass of bits of broken glass, odds and ends of brass, tin, wire, old bottles, burned corks, and broken instruments. Yet it was kaleidoscopic in its nature, and all resulted in effective and beautiful work. Experiments in the midst of this dusty medley formed the chosen and delightful occupation of his life. Writing was performed "doggedly" as the labour and the duty, but the long dark passages, the round hole or chink in the shutter, the ingeniously cobbled instrument, as well as his more elaborate telescopes and microscopes, formed the material of his greatest earthly enjoyment'[6].

Once settled at St. Andrews, he was able to use the resources of the natural philosophy class[7], amongst them extensive polarising apparatus 'with 16 attachments', and a microscope by Andrew Ross[8] to supercede his own, rather old-fashioned, anonymous instrument[9]. After the move to Edinburgh in 1859 as Principal of the University, he became a frequent visitor, despite his great age, to P. G. Tait's cluttered laboratory in the Old Quadrangle[10]. So, although it is impossible to estimate with any accuracy the quality of the instruments which were taken, at his own request, to the family home near Kingussie after his death[11], it may be deduced that as the best commercial scientific equipment was expensive, Brewster's own apparatus may not have been highly finished. Working with such equipment, Brewster nonetheless produced outstanding results.

References:
1. A. D. Morrison-Low, 'Brewster and Scientific Instruments', in this volume; the exhibition and this paper are mutually illustrative.
2. Science Museum, London: Science Museum Library, Archive Collection (Fox Talbot Papers). Letter dated 23 May 1838.
3. A. D. Morrison-Low, 'The Origins of the Polarising Microscope: Sir David Brewster *versus* William Nicol', (*forthcoming*).
4. See item 12 in this exhibition.
5. J. B. Morrell, 'Brewster and the Early British Association for the Advancement of Science', in this volume; J. Morrell & A. Thackray, *Gentlemen of Science: Early years of the British Association for the Advancement of Science* (Oxford, 1981), 345–346.
6. M. M. Gordon, *The Home Life of Sir David Brewster* (Edinburgh, 1869), 307–308.
7. [D. Smith], 'Articles in Sir D. Brewster's Classroom July 20, 1858'. (Manuscript in Hay Fleming Reference Library, St. Andrews).
8. See item 12 in this exhibition.
9. See item 17 in this exhibition.
10. C. G. Knott, *Life and Scientific Work of Peter Guthrie Tait* (Cambridge, 1911), 69.
11. Gordon, *op. cit.* (6), 414.

1. Reflecting telescope by James Veitch

Gregorian reflecting telescope, with metal specula, octagonal wooden barrel brass-bound at both ends, and pillar-mounted on a folding table tripod stand. The lower band is engraved 'JAMES VEITCH/INCHBONNY' (the other band is a replacement). The primary speculum is mounted within the lower band and is retained by a turned wooden endplate, threaded for an eyepiece (which is missing); the mirror surface is bright but pitted. The concave secondary speculum is mounted from an iron slide within the upper portion of the tube and is focused by an external brass rod threaded at its extremity. The knuckle joint on the pillar is geared for adjusting altitude (but lacks the worm drive).
Size: aperture 110 mm, focal length of primary about 460 mm. Overall length of tube 630 mm.

Provenance: From the Frank Collection of Scottish Scientific Instruments, and presented to the Royal Scottish Museum in 1981 (Inv. RSM TY 1981.37)[1]. Acquired by Mr. Frank privately in Glasgow, about 1975.

David Brewster's childhood in Jedburgh was strongly influenced by, amongst others, the self-educated mathematician and astronomer James Veitch of Inchbonny, and Brewster made his first telescope under Veitch's guidance at the early age of ten[2]. There is scant evidence of Veitch producing instruments commercially, but the family collection includes a number of pieces constructed by him and his son William[3]. This particular piece, and item 15 (Inv. RSM TY 1979.96) may however have been commissioned from Veitch or presented by him to friends[4].

References:
1. This piece, and James Veitch as an instrument-maker, will be discussed further in T. N. Clarke, *Scottish Scientific Instruments from the Frank Collection in the Royal Scottish Museum (forthcoming)*.
2. M. M. Gordon, *The Home Life of Sir David Brewster* (Edinburgh, 1869), 22–148, especially 30.
3. Now exhibited at the Royal Observatory, Edinburgh. Items not restricted by entail were sold at Messrs. Christie, Wednesday, 9 April 1975.
4. One such piece was commissioned through Brewster in 1821 for Professor Heinrich Schumacher, at Altona Observatory. This was '... an order for one of his best telescopes, a Gregorian reflector ... 2 feet 8 in focal length, and 5 inches aperture and proved a very fine instrument. A telescope of the same description was made for the Earl of Minto, and another is in the possession of his son, the Rev. Dr. Veitch, at Merchiston, near Edinburgh'. Gordon, *op. cit.* (2), 102.

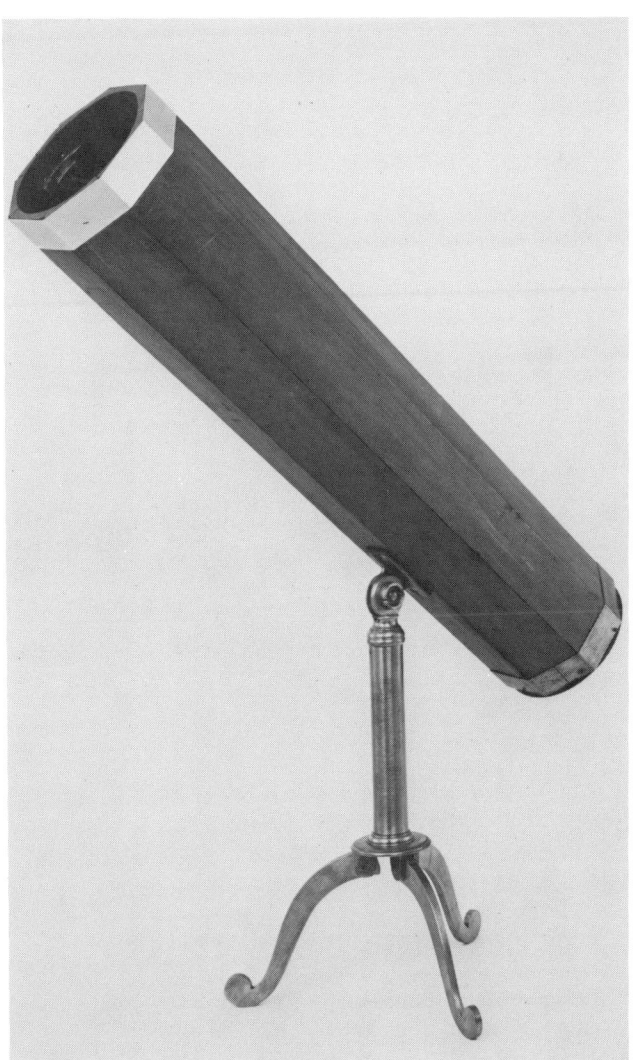

Fig. 11. *Cat. 1.* Reflecting telescope by James Veitch.

2. **David Brewster** *A Treatise on New Philosophical Instruments*.
Leather-bound volume, entitled *A Treatise on New Philosophical Instruments, for various purposes in the Arts and Sciences. With Experiments on Light and Colours*. By David Brewster LL.D. Fellow of the Royal Society of Edinburgh, and of the Society of Antiquaries of Scotland. (Edinburgh: Printed for John Murray, Albemarle Street, London; and William Blackwood Edinburgh, 1813). Quarto volume of 427 pages printed by A. Balfour, in Edinburgh, and 12 plates engraved by J. Moffat.
Size: 150 × 230 × 50 mm.

Provenance: Purchased by the Royal Scottish Museum Library in 1977.

Brewster's first book appeared in 1813, and was devoted to descriptions of scientific instruments and his determinations of the refractive and dispersive powers of nearly two hundred substances made during his attempts to improve achromatic telescopes[1]. As Morse explains, news of similar work on the Continent moved his interest from instruments back to optical theory; however, improvement of instrumentation continued to hold a lifelong fascination for him[2].

References:
1. E. W. Morse, 'Sir David Brewster', in C. C. Gillispie (ed.) *Dictionary of Scientific Biography* 16 vols (New York, 1970–76) II, 451.
2. A. D. Morrison-Low, 'Brewster and Scientific Instruments', in this volume.

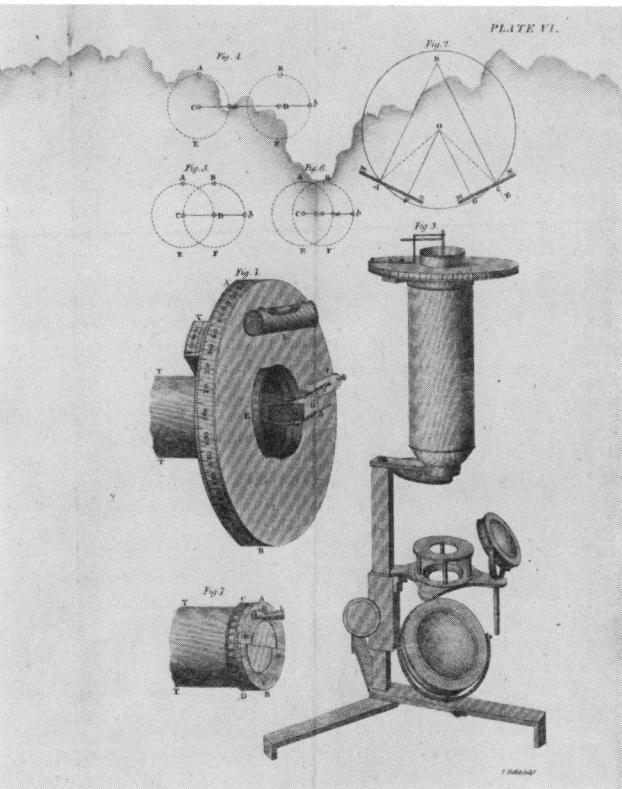

Fig. 12. *Cat. 2.* Plate VI from *A Treatise on New Philosophical Instruments* (1813).

3. **Telescope for measuring angles and distances by William Harris**
2 inch (48 mm) aperture three draw non-achromatic telescope in wood and brass. The innermost tube engraved 'W. Harris & Cº/50 HOLBORN LONDON/PATENT TELESCOPE/ for Meafuring Angles & Diftances/Nº 53'. The middle draw with a fixed separation divided lens at the objective end and engraved along the tube a 'Scale of

Minutes Divided object Glafs' (numbered in tens from 40 to 60 and divided to 2″ arc) and a 'Scale of Minutes Single object Glafs' (numbered in tens from 170 to 210 and divided to 3″ arc), the two scales read against the flange of the outer draw tube. Objective cover engraved '64' and threaded to a brass sleeve with a modern marking 'N.S.L./V.3305', and possibly from another instrument[1].
Size when closed: 320 × 60 mm diameter.

Provenance: Lent by Thomas H. Court[2], in 1913, later presented to the Science Museum (Inv. Sc. M. 1913–288).

Brewster first described this instrument, among many others, in his first book published in 1813[3]. Two years earlier, he and a London scientific instrument maker, William Harris[4], had taken out a patent for this telescope[5]. This seems to be the only example in a public collection.

References:
1. Correspondence from the Science Museum, August 1982 'In October 1922 Mr. Court added a dividing lens, and replaced the existing cap with "a proper brass cap"'.
2. For an assessment of Thomas Court's role as a collector of scientific instruments, see Harold Heywood, 'Clay and Court—Pioneer Collectors of Microscopes', *Microscopy*, 31 (1969), 109–112; and Jane Insley, 'Court, Crisp and Clay—some Notes on Collectors and Collections of Antique Microscopes', *Microscopy*, 34 (1982), 345–353, 376.
3. D. Brewster, *Treatise on New Philosophical Instruments* (Edinburgh, 1813), 173–191: item 2 in this exhibition.
4. E. G. R. Taylor, *Mathematical Pactitioners of Hanoverian England* (Cambridge, 1966), 365, entry 1151 (Harris); 357, entry 1088 (Brewster)
5. Patent Specification 1811, No. 3453.

Fig. 13. *Cat. 3.* Telescope patented by D. Brewster and W. Harris in 1811.

4. Patent documents for the kaleidoscope
(a) *Specification*
'Dated 27 Aug 1817/Specification of David Brewster LLD./Inrolled upon The Roll of Specifications and Surrounders in the Chapel of the Roll of the High Court of Chantry the thirtieth day of August in the fifty-seventh year of the reign of His Majesty King George the Third and in the year of Our Lord one thousand eight hundred and seventeen (signed) John Kipling'. [Clerk].
Size: 650 × 515 mm.

(b) *Letters patent*
Printed sheet of parchment, with the specification written in full. Great Seal of the United Kingdom, in red wax protected in metal container, attached to document by a silken cord.
Size: 660 × 540 mm.
Seal size: 170 mm diameter.

(c) *Legal signatures* [not displayed or illustrated]
Signed by Robert and Archibald Montgomery. [Lawyers].
Size: 205 × 320 mm.

All three papers contained in a leather covered box with gold tooling.
Size: 300 × 235 × 75 mm.

Provenance: Lent by Mrs. M. Brewster-Macpherson.

The patent documents for the kaleidoscope[1] have been held by the descendants of Brewster's family since they were first granted[2]. Brewster himself held high hopes for the instrument[3], writing an instruction manual issued with certain examples[4], an extended treatise[5], an encyclopaedia article[6] and a history[7]. However, despite the protection of a patent, Brewster felt that the system had let him down and he had lost a potential fortune[8].

References:
1. Listed with other family papers in N.R.A. (Scot.) Survey 0718.
2. Patent Specification 1817, No. 4136.
3. M. M. Gordon, *The Home Life of Sir David Brewster* (Edinburgh, 1869), 95–99.
4. D. Brewster, *Description and Method of Use of the Patent Kaleidoscope* (London, 1818).
5. D. Brewster, *A Treatise on the Kaleidoscope* (Edinburgh, 1819).
6. [D. Brewster], 'Kaleidoscope', in *Edinburgh Encyclopaedia* 20 vols. (Edinburgh, 1830) XII, 410–412.
7. D. Brewster, *The Kaleidoscope: its History, Theory and Construction with its Application to the Fine and Useful Arts* (Edinburgh, 1858).
8. A. D. Morrison-Low, 'Brewster and Scientific Instruments', in this volume.

Fig. 14. *Cat. 4.* Patent documents for the kaleidoscope.

5. Polyangular Kaleidoscope by R. B. Bate
Polyangular kaleidoscope in brass, marked under the supporting collar 'Bate London' and '4/6/8/10/12/16/20/30/40/60//18/24/36/48'; also engraved on the body tube above the collar with The Royal Coat of

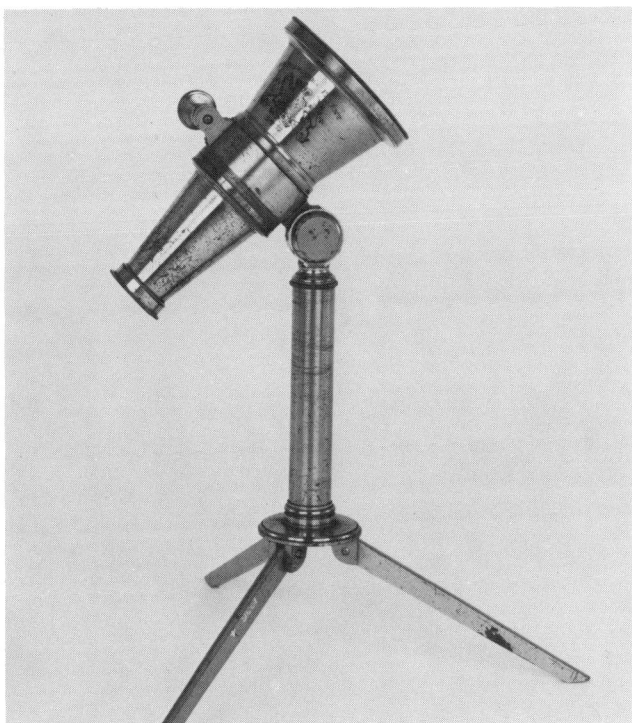

Fig. 15. *Cat. 5.* Polyangular kaleidoscope by R. B. Bate.

Fig. 16. *Cat. 6.* Telescopic kaleidoscope by W. Harris & Co.

Arms and 'D^r. BREWSTER'S PATENT/N° 37'. When the collar is slackened by turning the milled screw, the body of the instrument can be turned within, so that the internal polished mirrors are moved, reflecting the number of images marked under the collar, eg. at mark '4' the mirrors are set at 90°, at mark '6', mirrors set at 60° etc. The cone-shaped body tube, is mounted on a trunion set at the top of a pillar, which is screwed into a folding tripod stand.
Size of body: 135 mm × 73 mm diameter.
Instrument: 300 × 200 × 200 mm.

Provenance: Presented by Dr. Margaret Dobson to the Royal Scottish Museum in 1936 (Inv. RSM TY 1936.4).

The instrument was initially offered to the Science Museum, London, before its acquisition by the Royal Scottish Museum[1]. Other examples of Robert Brettell Bates' instrument in public collections in the United Kingdom are numbered as follows: Science Museum, London: Nos. unknown[2] and 45[3]; Whipple Museum of the History of Science, Cambridge: Nos. 30[4] and 40[5]. Bate[6] does not appear to have made more than about sixty numbered examples; by the 1820s he was greatly involved in the making of standard measures for the Board of Trade.

References:
1. RSM Technology Department, correspondence files, letter of 24 February 1936.
2. Inventory number: 1918–113. Lent by Thomas H. Court, returned in 1927.
3. Inventory number: 1928–881.
4. Inventory number: 601 (without stand).
5. Inventory number: 601 (with stand).
6. E. G. R. Taylor, *The Mathematical Practitioners of Hanoverian England* (Cambridge, 1966), 355, entry 1079.

6. Telescopic Kaleidoscope by William Harris & Co
Two draw telescopic kaleidoscope in brass and lacquered tin, the eyepiece embossed 'W. HARRIS & Co/50 HOLBORN, LONDON'. The folding brass tripod is detachable by unscrewing a collar; it dismantles and fits into the box. A trade card (partly defaced by handwriting) reads 'WILLIAM HARRIS & Co/Manufacturers of/Optical,

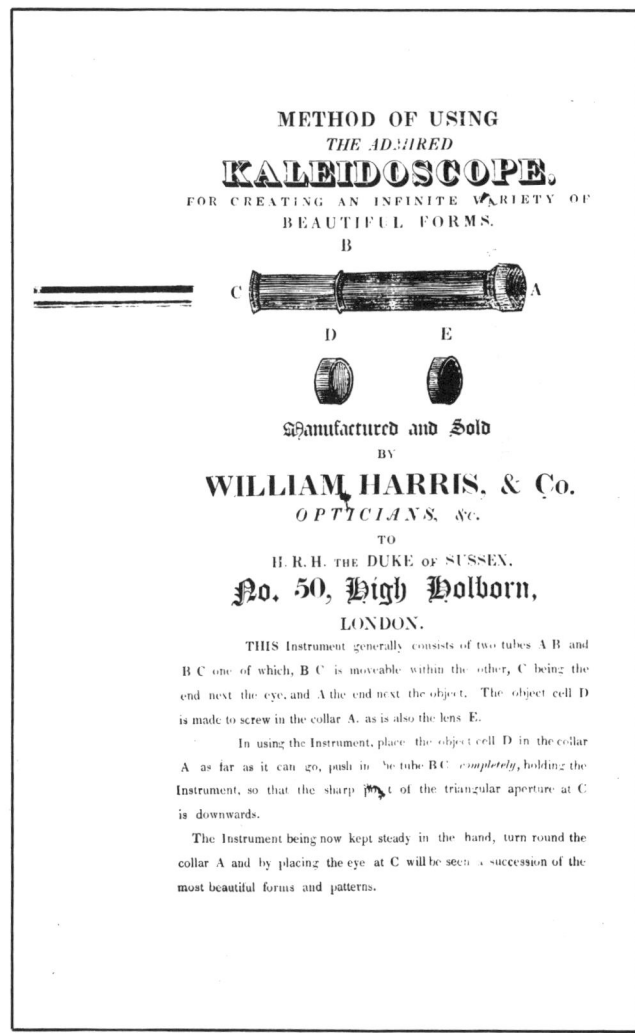

Fig. 17. *Cat. 6.* Instruction sheet for telescopic kaleidoscope.

Mathematical and Philosophical/INSTRUMENTS/50 High Holborn, Corner of Brownlow Street', and an octavo printed instruction sheet headed 'METHOD OF USING/THE ADMIRED/KALEIDOSCOPE', is also enclosed. There are also coloured glass pieces to put in

circular cells, and five painted glass roundels in a cylindrical wooden holder.
Size, in use, unextended: 450 mm high × 240 × 240 mm.
Tube length: 240 × 50 mm diameter.
Box: 290 × 140 × 65 mm.

Provenance: Presented by Miss C. Moody to the Science Museum, London, in 1931 (Inv. Sc. M. 1931–668).

William Harris, who had earlier taken out a joint patent with Brewster[1], does not appear in Brewster's list of approved kaleidoscope retailers[2]; but then Brewster's name does not appear in the instruction leaflet issued with this instrument[3]. It is argued elsewhere that Harris had bungled the patenting of the kaleidoscope[4].

References:
1. See item 3.
2. [D. Brewster], 'Kaleidoscope', in *Edinburgh Encyclopaedia* 20 vols. (Edinburgh, 1830), XII, 412.
3. *Method of Using the admired Kaleidoscope, for creating an infinite variety of beautiful forms. Manufactured and sold by William Harris. & Co. Opticians etc.*
4. A. D. Morrison-Low, 'Brewster and Scientific Instruments', in this volume.

Fig. 18. *Cat. 7.* Telescopic kaleidoscope by Carpenter.

7. Telescopic Kaleidoscope by Philip Carpenter

Two draw telescopic kaleidoscope with crimson enamelled body tube, the eyepiece embossed 'P. CARPENTER/SOLE MAKER'. On brass aperture end, stamped with heraldic device—lion and unicorn with Garter between them, 'HONI SOIT QUI MAL Y [PENSE]' and 'DIEU/ET/MON/DROIT'—above aperture, and stamped 'CARPENTER &/WESTLEY/KALEIDOSCOPE' below. Two push-fit colour glass cells, with eleven others in box. The box is lined with green silk and green velvet, and its lid has an inlaid oval stamped metal label 'DR. BREWSTER'S/PATENT/KALEIDOSCOPE'.
Size: unextended: 165 × 35 mm diameter.
Box: 185 × 95 × 50 mm.

Provenance: Lent by Thomas H. Court to the Science Museum, London, in 1918; later presented (Inv. Sc. M. 1918–112).

Philip Carpenter, unlike William Harris, seems to have remained in favour with Brewster, his name appearing on the list of approved kaleidoscope retailers[1]. It would appear that Carpenter continued to make kaleidoscopes long after the business moved from Birmingham[2] to London[3] as is illustrated by this particular instrument, stamped 'CARPENTER &/WESTLEY', the name to which the firm changed in 1837[4].

References:
1. See Item 6 in this exhibition.
2. S. Timmins (ed.), *The Resources, Products and Industrial History of Brimingham* (London, 1866), 534. I am grateful to D. J. Bryden for this reference.
3. R. H. Nuttall, 'Philip Carpenter and the "Microcosm" Exhibition', *Microscopy*, 33 (1966), 62–65.

4. An earlier instrument in the Science Museum London (Inv. Sc. M. 1911–270), is closely similar but has brass body work and black enamelled body tube, and instead of 'CARPENTER &/WESTLEY' is stamped 'DR. BREWSTER'S/PATENT/KALEIDOSCOPE'. This example is housed in a cylindrical cardboard case, and the instrument's size unextended is 170 × 35 mm diameter. It was presented by Thomas Court in 1911.

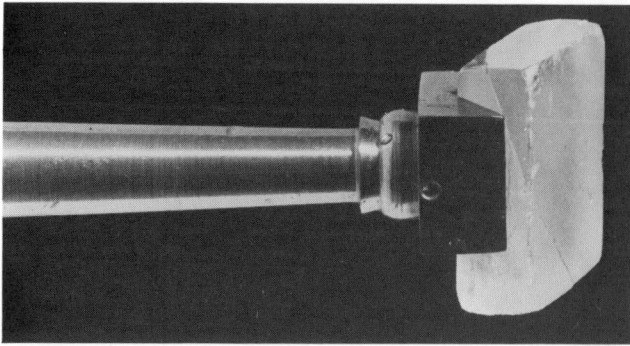

Fig. 19. *Cat. 8.* Nicol prism made by William Nicol.

8. Nicol prism made by William Nicol

Nicol prism of Iceland spar, supported on a brass pillar. The Canada balsam has crystallised with age, and the prism is almost opaque.
Size: 16 × 7 × 7 mm.
Pillar height: 55 mm.

Provenance: RSM Register 1856 'June 3rd. Analysing Prism of Calespar (Nicols') made by the inventor the late William Nicol in his 80th year. Presented by Alexander Bryson, 66 Princes Street Edinburgh.' (Inv. RSM 1856.54).

William Nicol[1] was a lecturer on natural philosophy in Edinburgh[2], but does not appear to have been formally attached to the University. A list of his published papers[3] shows that his interests were predominantly practical in the fields of mineralogy and microscopic techniques: experimenting with polarising light appears to have been incidental to these pursuits. The paper describing his prism[4] appeared in 1829 when he was aged about 61, and it appeared in a journal edited by Robert Jameson[5] rather than in Brewster's *Edinburgh Journal of Science*[6]. He was to improve his invention some ten years later[7], and after his death he left his 'Books, Philosophical Apparatus, Minerals, Fossils, Shells, Optical Instruments' including his prisms[8] to Alexander Bryson[9], who was a friend of the first Director of the Industrial Museum of Scotland, George Wilson. In 1856 Wilson was actively looking for suitable material for the new museum, now the Royal Scottish Museum[10]. Bryson and Nicol's joint collections were sold on Bryson's death[11]; part of these are now in the Mineralogy Department of the British Museum (Natural History)[12]. James Clerk Maxwell once owned a pair of Nicol prisms, made by Nicol himself[13]. Another prism, possibly by Nicol, is in the Museum of the History of Science, Oxford[14].

References
1. E. Frankel, in C. C. Gillispie (ed.), *Dictionary of Scientific Biography* 16 vols. (New York, 1970–76) X, 109–110; this entry illustrates how little is known about Nicol.
2. This is how he was described from 1814: *Memoirs of the Wernerian Natural History Society*, 2, Appendix 'History of the Society', 653.
3. *Royal Society Catalogue of Scientific Papers* (1800–1863) (London, 1867–72) IV, 615–616.
4. W. Nicol, 'On a Method of so far Increasing the Divergence of the Two Rays in Calcareous Spar, that Only One Image may be seen at a Time', *Edinburgh New Philosophical Journal*, 7 (1829), 83–84.

5. J. M. Eyles, in C. C. Gillispie (ed.), *Dictionary of Scientific Biography* 16 vols. (New York, 1970–76) VII, 69–71.
6. W. H. Brock, 'Brewster as a Scientific Journalist', in this volume.
7. W. Nicol, 'Notice Concerning an Improvement in the Construction of the Simple Vision Prism of Calcareous Spar', *Edinburgh New Philosophical Journal*, 27 (1839), 332–333.
8. William Nicol's Will. Scottish Record Office. SC70/4/17, 959.
9. Robert Bryson and his sons James and Alexander are discussed in T. N. Clarke, *Scottish Scientific Instruments from the Frank Collection at the Royal Scottish Museum* (forthcoming).
10. R. G. W. Anderson, *The Playfair Collection, and the Teaching of Chemistry at the University of Edinburgh 1713–1858* (Edinburgh, 1978), 57.
11. J. M. Chalmers-Hunt, *Natural History Auctions 1700–1972* (London, 1976), 106: 10 March 1868, sold by Stevens Auction Rooms, London.
12. W. N. Edwards, 'William Nicol and Henry Clifton Sorby' *Nature*, 168 (1951), 566–567.
13. Richard Glazebrook, *James Clerk Maxwell and Modern Physics* (London, 1893), 20–21; these were then in the Cavendish Laboratory, Cambridge.
14. Correspondence from the Museum of the History of Science, Oxford, December 1982: this prism probably came from Bryson's shop *via* the Clay Collection.

Fig. 20. *Cat. 9.* Rhomb of Iceland spar.

9. Rhomb of Iceland spar
Large cleavage rhomb of the trigonal crystal calcite.
Size: 120 × 70 × 60 mm.

Provenance: Presented by the late John R. Hutchison to the Royal Scottish Museum (Inv. RSM TY 1967.181), and originally lent by Mr. Hutchison to the Museum in 1959.

The doubly refracting qualities of Iceland spar first came to the attention of natural philosophers in the seventeenth century[1], but it was only useful as an experimental tool after William Nicol's invention of his prism[2]. There is no correspondence concerning this particular piece of calcareous spar in the Royal Scottish Museum archives, and its history, if any, is unknown. The optician, John R. Hutchison, was a son of George Hutchison, who had previously worked for the instrument-makers Adie & Wedderburn, successors to Alexander and John Adie, and continued their business[3].

References:
1. F. C. Cheshire, 'The President's Address: The Early History of the Polariscope and the Polarizing Microscope', *Journal of the Royal Microscopical Society*, 43 (1923), 1–18.
2. Cheshire, *op. cit.* (1); F. C. Cheshire, 'Polariscopes: a few Typical Forms of Early Instruments in the South Kensington Museum', *Transactions of the Optical Society*, 23 (1921–22), 246–255.
3. D. J. Bryden, *Scottish Scientific Instrument Makers 1600–1900* (Edinburgh, 1972), 27 n. 132.

Fig. 21. *Cat. 10.* Nicol prisms used by W. H. F. Talbot.

10. Nicol prisms used by W. H. F. Talbot
(a) Nicol prism mounted in two cork half cylinders, the two ends and internal surfaces of the cork stained with black ink. The two pieces of cork are glued together to form a cylinder with the prism contained between them.
Size (total): 48 × 32 mm diameter.
(prism only): 40 × 12 × 12 mm.

(b) Nicol prism as above, the whole in a cylindrical brass mount, with both ends screw-topped. No maker's name.
Size (total): 30 × 22 mm diameter.
(prism only): 20 × 6 × 6 mm.

Provenance: RSM Register 1936 'Prism, Nicol, $1\frac{1}{4}''$ dia. × $1\frac{7}{8}''$ long Prism, Nicol, $\frac{23}{32}''$ diameter × $\frac{31}{32}''$ long in brass case with screw caps on ends. Presented by Miss M. Talbot, Lacock Abbey, Chippenham, Wilts'. (Inv. RSM TY 1936.106 and 1936.108).

In 1936 Mr. Alexander Barclay of the Science Museum began a new file on what was to become a major accession to that institution[1]. Miss Matilda Talbot[2], grand-daughter of the Victorian polymath William Henry Fox Talbot, had decided to entrust a large portion of what survived of her grandfather's apparatus and papers to three national institutions—The Science Museum, the Royal Photographic Society and the Royal Scottish Museum. The major part of this historic collection arrived in these institutions before the outbreak of the Second World War. W. H. F. Talbot[3] is usually remembered as the inventor of positive/negative photography[4], but he was also a man of science, with interests ranging from optics to Egyptian hieroglyphics[5]. He first read about Nicol's invention in a German scientific journal[6], and not, surprisingly, from his Scottish friend and correspondent Sir David Brewster. Further papers[7] by Talbot led Brewster to credit him with the invention of the polarising microscope[8]. These prisms may have been made by Watkins of Charing Cross, whose polarisation apparatus Talbot recommended[9].

References:
1. J. P. Ward, 'The Fox Talbot Collection at the Science Museum', *History of Photography*, 1 (1977), 275–287.
2. Matilda Talbot, *My Life at Lacock Abbey* (London, 1956), *passim.*
3. R. V. Jenkins in C. C. Gillispie (ed.), *Dictionary of Scientific Biography* 16 vols. (New York, 1970–76) XIII, 237–239.
4. D. B. Thomas, *The First Negatives* (London, 1964).
5. H. J. P. Arnold, *William Henry Fox Talbot: Pioneer of Photography and Man of Science* (London, 1977).
6. [W.] H. F. Talbot, 'On Mr Nicol's Polarizing Eyepiece', *London and Edinburgh Journal of Science and Philosophical Magazine*, 3rd series 4 (1834), 289–290.
7. [W.] H. F. Talbot, 'Microscopic Appearances with Polarized

Light', *London and Edinburgh Journal of Science and Philosophical Magazine*, 3rd series 5 (1834), 321–327; W. H. F. Talbot, 'On some Optical Experiments', *Proceedings of The Royal Society of Edinburgh, 7* (1872), 466–470.
8. D[avid] B[rewster], 'Microscope' in *Encyclopaedia Britannica*, 22 vols. 8th edition, (Edinburgh, 1856), 789.
9. Talbot, *op. cit.* (6), 290n.

Fig. 22. *Cat. 11.* Two tourmaline crystals, used by Sir David Brewster.

11. Two tourmaline crystals, used by Sir David Brewster

Two thin sections of tourmaline each mounted between glass in two circular brass holders with attachments for a small optical bench (now missing). In cardboard box, marked 'Pair of Tourmalines'. One has lost half its protective glass.
Size: 75 mm long × 23 mm diameter (tourmaline mounts) × 15 mm diameter (mounting boss).

Provenance: Lent by the Department of Physics, St. Andrews University.

These two pieces appear to be part of a piece of optical apparatus lent to Sir David Brewster while he was Principal of the United Colleges of St. Salvator and St. Leonard at St. Andrews, and returned by him with another ten pieces (including item 12 of this exhibition) after he moved to Edinburgh[1]. The entry reads:

'5. Small Apparatus in brass stand for showing rings in Crystals with two Tourma-lines, two lenses and eleven crystals'.

The rest of this apparatus has not been identified. Brewster's investigations of the polarisation of light began as early as 1813[2] and he later claimed anonymously that 'the polarizers and analyzers which he used were bundles of glass plates, either reflecting or transmitting light, tourmalines, or small rhombs, or achromatised prisms of calcareous spar'[3].

References:
1. 'Apparatus Committee's Report', 16 March 1861 (St. Andrews University Library, Department of Manuscripts), 143.
2. D. Brewster, 'On some Properties of Light', *Philosophical Transactions of the Royal Society of London, 103* (1813), 101–109.
3. [David Brewster], 'The Microscope and its Revelations', *North British Review, 25* (1856), 464.

12. Microscope by Andrew Ross

Monocular Lister-limb microscope in brass, engraved 'AND^W. ROSS & C^O./Opticians/33 Regent S^t. Piccadilly' on one arm of the tripod base. A pillar supports the Lister-limb on a trunion, which has an adjusting clamp screw. Coarse focus is by rack and pinion on the limb; fine focus by a milled screw acting on a lever operating on the nosepiece. Mechanical stage with centering substage. Four objectives,

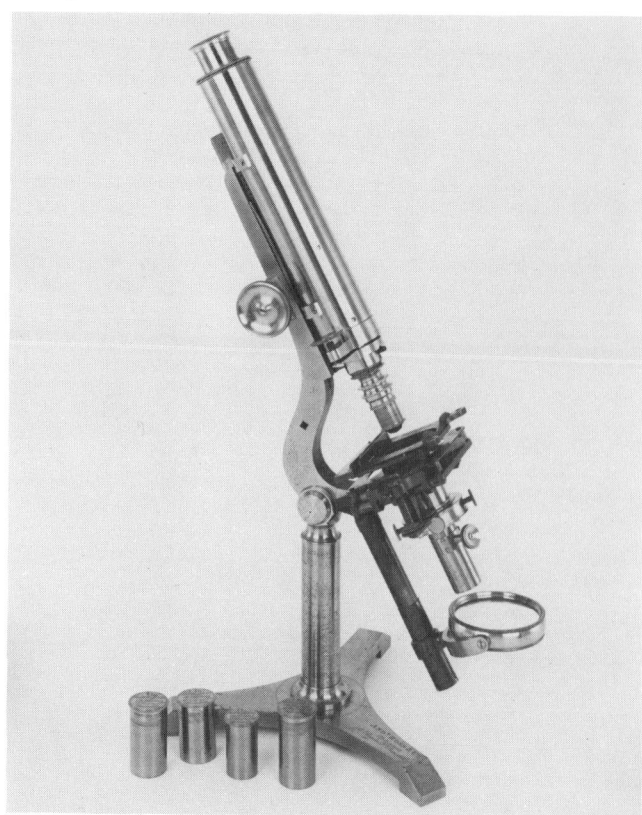

Fig. 23. *Cat. 12.* Compound microscope by Andrew Ross.

$1''$, $\frac{1}{2}''$, $\frac{1}{4}''$ and $\frac{1}{8}''$ otherwise unmarked, all in brass capsules marked "And^w. Rofs & Co./Opticians/33 Regent S^t./Piccadilly', with three eyepieces.
Size (in use): 210 × 210 × 420 mm.
Box: 275 × 260 × 465 mm.

Provenance: Lent by the Department of Physics, St. Andrews University.

In 1861 this microscope was returned by Brewster[1] after he moved from St. Andrews to Edinburgh[2], along with another ten pieces (including item 11 of this exhibition). The entry reads:

'No. 1 Microscope by Ross. Powers 1 inch/$\frac{1}{2}$, $\frac{1}{4}$, $\frac{1}{8}$— Polarizing Apparatus &c.'

This Lister-limb instrument can be dated between 1838 and 1842[3] from the style and manner of the engraved signature; and more precisely its purchase is recorded as follows:

'Paid from app^s fund to Sir D. Brewster . . . 1840 July 28 for a ~~telescope~~ microscope 52.8.8'[4].

Another list notes '1840 Ap. 19 Ross 52.8.8'[5], and further states that other apparatus is 'Taken up to college from Sir D^{ds} House, . . . the Microscope being with him' on July 22nd 1858. Presumably he was still using it.

Certainly, Brewster had nothing but praise for Ross—'the microscope of Mr. Ross must be regarded as the finest hitherto constructed'[6],—and as he used it consistently between 1840 and 1858, it must have reached his exacting standards[7].

References:
1. 'Apparatus Committee's Report', 16 March 1861 (St. Andrews University Library, Department of Manuscripts), 143.
2. M. M. Gordon, *The Home Life of Sir David Brewster* (Edinburgh, 1869), 288.
3. A. D. Morrison-Low and R. H. Nuttall, 'Ross Microscopes as

used by David Brewster and Richard Owen', *Microscopy*, 34 (1982), 335–344.
4. D. Smith, 'Scroll Copy. Inventory of the Apparatus in Nat. Philosophy Class of the United College May 1847', (Manuscript in Hay Fleming Reference Library, St. Andrews).
5. [D. Smith], 'Apparatus Sir D. Brewster. July 1858', (Manuscript in Hay Fleming Reference Library, St. Andrews).
6. For instance, D[avid] B[rewster], 'Microscope' in *Encyclopaedia Britannica* 22 vols. 8th edition, (Edinburgh, 1856), 781.
7. David Brewster, *Treatise on Optics* (London, 1853), 475–476: Brewster discusses his own 'additional pieces for a fine microscope by Ross, which I have long used'.

Fig. 24. *Cat. 13.* Polarising apparatus by Bryson, Edinburgh.

13. Polarising apparatus by James Bryson

Boxed apparatus consisting of black reflecting glass contained inside the lid, Nicol prism mounted on adjustable brass pillar hinged inside outer edge of box, and glass stand hinged to inside inner box edge, tilting to allow transparent objects placed on the glass to be illuminated by light partially polarised by inclined reflection from the black glass surface. The Nicol prism acts as an analyser, revealing patterns of stress in the 17 differently shaped glass pieces and colours in the 3 selenite designs placed on the glass stand. Glass stand engraved by the hinge 'Bryson. Edinburgh'.
Box size (shut): 222 × 160 × 45 mm.
(in use): 444 × 160 × 21 mm.

Provenance: RSM Register 1860 'February 17th. Apparatus for polarizing light. 17 pieces of unannealed glass and three Selenite designs for polarizer. Bought from James Bryson'. (Inv. RSM 1860.525).

This apparatus appears to be a demonstration piece, showing the lines of strain in the unannealed glass shapes. Educational apparatus similar to this was still being sold in the 1930s[1]. Brewster discovered that transparent isotropic materials become optically anisotropic when subject to mechanical stress in 1816[2], and this photo-elastic effect is used today by engineers to investigate stresses in structures where calculation would be very laborious. A model of the structure is made in a suitable transparent plastic, to which stress is gradually applied[3]. Brewster's initial discoveries were enlarged upon by James Clerk Maxwell[4], but the phenomenon did not become widely investigated until the twentieth century, by which time its origins appear to have been forgotten[5].

The maker of this piece, James Bryson, acquired some of William Nicol's prisms for his Princes Street Shop from his brother Alexander, who had inherited them on Nicol's death in 1851[6].

References:
1. Advertisement in *Watson's Microscope Record* No. 29 (May 1933), 21.
2. D. Brewster, 'On the Communications of the Structure of Doubly-refracting crystals to Glass, Muriate of Soda, Fluor Spar, and Other Substances by Mechanical Compression and Dilatation', *Philosophical Transactions of the Royal Society of London*, 106 (1816), 157–178.
3. R. W. Ditchburn, *Light* 2nd edition (London and Glasgow, 1963), 642–643.
4. J. P. M. Pannell, *An Illustrated History of Civil Engineering* (London, 1964), 336–337.
5. For instance, M. Focht (ed.), *Photoelasticity: Proceedings of the International Symposium held at Illinois Institute of Technology October 1961* (Oxford, London, New York, Paris, 1963), vii, states 'Although technical interest in photoelasticity dates back to the turn of the century when Mesnager in France, and Coker and Filon in England did their pioneer work, conferences on the subject did not begin before 1930'.
6. D. J. Bryden, *Scottish Scientific Instrument Makers 1600–1900* (Edinburgh, 1972), 45. James MacKay Bryson is discussed in T. N. Clarke, *Scottish Instruments from the Frank Collection at the Royal Scottish Museum (forthcoming)*.

Fig. 25. *Cat. 14.* Pill-box with garnets.

14. Pill-Box with Garnets

Oval box made from thinly shaved wood, the lid marked in pencil 'Garnets'. Inside, a piece of crumpled paper 85 × 70 mm marked in David Brewster's handwriting 'From Dr. Brewster' and on the other side 'Precious Garnet'. Inside a piece of tissue paper are some thirty one small pieces of uncut material, with two clear and two coloured pieces which have been faceted, (the latter being originally costume jewelry).
Size of box: 73 × 40 × 30 mm.

Provenance: Lent by W. H. Veitch Esq., of Inchbonny.

The box remained with the Veitch family at Inchbonny after David Brewster sent it to James Veitch[1], and it is presumed that this is the remainder of the sample supplied in 1828 from which it was intended Veitch would grind jewel lenses[2]. Brewster's friendship with Veitch[3], begun during his childhood, continued through his undergraduate days at Edinburgh University[4] and only ended with the death of the older man in 1838[5].

References:
1. NRA(Scot) Survey/0337/63 lists a letter from David Brewster to James Veitch dated 12 October 1828, with which Brewster sends fragments of garnet 'sufficiently large for your purpose'.
2. M. M. Gordon, *The Home Life of Sir David Brewster* (Edinburgh, 1869), 102; R. H. Nuttall & A. Frank, 'Makers of Jewel Lenses in Scotland in the Early Nineteenth Century', *Annals of Science*, 30 (1973), 410.
3. Discussed with item 1 in this exhibition.
4. W. Cochran, 'Sir David Brewster: An Outline Biography', in this volume.
5. Gordon, *op. cit.* (2), 140.

Fig. 26. *Cat. 15.* Simple microscope by James Veitch.

15. Simple microscope by James Veitch

Amateur-made simple microscope in brass, mounted on a brass shoe stamped 'JAMES VEITCH/INCHBONNY/ 1834' attached to sliding wooden lid of box. The instrument consists of two parallel brass rods, one of which is fixed to a rectangular brass base at the foot and has the eyepiece holder at the top, the rectangular wooden-framed mirror being positioned between these. The other brass rod is fixed at the bottom to a sliding brass piece moved by turning screw, forming the focusing mechanism, and is attached to the double-jointed swing stage at the top. The stage is circular, with a glass insert; the single simple lens held in a removable sliding brass mount, is also made of glass. Size: 118 × 56 × 150 mm.

Provenance: Purchased by The Royal Scottish Museum in 1979[1] (Inv. RSM TY 1979.96).

Like the telescope made by James Veitch[2] this microscope is one of his few known pieces made for donation or sale. Its construction is similar to that of two undated simple microscopes by him which remain in the family collection. Although the lens is made from glass[3], the instrument is probably of the type for which garnet lenses were intended[4].

References:
1. Norman Tebble (ed.), *Royal Scottish Museum Triennial Report 1977–78–79* (Edinburgh, 1979), 51.
2. Item 1 in this exhibition, RSM TY 1981.37.
3. Examined by Mr. R. J. Reekie, Department of Geology, Royal Scottish Museum, November 1980.
4. Item 14 in this exhibition.

16. Simple microscope

Amateur-made simple glass lens in brass holder. Two strips of sheet brass are rivetted together to form a handle; further along, one piece widens out into a larger rectangle surrounding the lens, which is held between the two strips. A third, smaller, rectangular piece is wedged between the

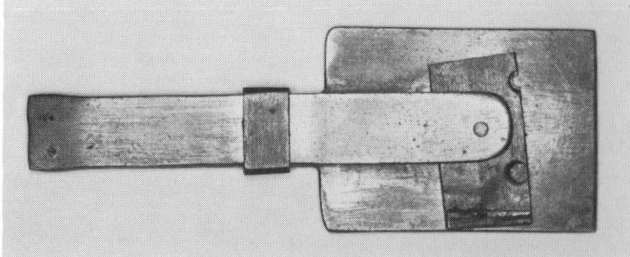

Fig. 27. *Cat. 16.* Simple microscope, once owned by Brewster.

two, with a hole for the lens punched in it; two other attempts to punch holes are visible. A fourth piece of brass holds the first two together, in a band around the handle. Size 53 × 19 mm.

Provenance: Presented by Thomas H. Court to the Science Museum, London. He had acquired it from the Crisp Collection[1] at one of the four sales during 1920 and 1921. (Inv. Sc.M. 1921–773). Its entry in the manuscript Crisp Catalogue[2] reads: 'DESIGNER AND MAKER: Sir David Brewster. HISTORY: Given me in 1886 by Mr. J. Mayall Esq., to whom it was given by Sir D. Brewster about 1861'[3].

Brewster had made scientific instruments from an early age, initially under the supervision of James Veitch of Inchbonny[4]. His interest in optics, especially microscopy, took a practical as well as theoretical turn: he constructed single[5], doublet and triplet[6] fluid lenses for microscopes, and achromatic doublets and triplets 'by combining a fluid *concave* lens with one or two convex lenses of precious stones or glass'[7], besides making what his daughter called 'the ingeniously cobbled instrument'[8]. The above piece would certainly be within his capabilities.

References:
1. Hugh B. C. Pollard, 'The Evolution of the Microscope', *The Illustrated London News* (14 February 1925), 250–252; Jane Insley, 'Court, Crisp and Clay—some notes on Collectors and Collections of Antique Microscopes', *Microscopy*, 34 (1982), 345–353, 376.
2. In the Court Papers, Science Museum Library, London. Sir Frank Crisp Bt. was Hon. Secretary, Royal Microscopical Society 1878–1889: *Who was Who*, Vol. 2. *1916–1928* (London, 1967), 245.
3. John Mayall was a member of the Royal Microscopical Society and an active collector of microscopes for both the Crisp and the Billings Collections: Insley, *op. cit.* (1); R. S. Clay and T. H. Court, *The History of the Microscope* (London, 1932), v, n.; and Helen R. Purtle (ed.), *The Billings Microscope Collection of the Medical Museum Armed Forces Institute of Pathology* (Washington D.C., 1974), v. Mayall delivered the Cantor Lectures on 'The Microscope', which later appeared in *Journal of the Society of Arts* (1886).
4. M. M. Gordon, *The Home Life of Sir David Brewster* (Edinburgh, 1869), 22–140, *passim.*; see items 1 and 15 in this exhibition.
5. D[avid] B[rewster], 'Microscope' in *Encyclopaedia Britannica* 22 vols. 8th edition, (Edinburgh, 1856), 768.
6. *Ibid.*, 775.
7. *Ibid.*
8. Gordon, *op. cit.*, (4), 308.

17. Chest microscope with garnet lens

Portable chest microscope in wood and brass, with square sectioned pillar hinged to a trunion joint inside the box. The pillar may be raised, but no further than illustrated. The body tube is screwed into a ring on an arm at the top of the pillar: focusing is by moving the stage on the pillar by rack and pinion. The concave mirror is fixed at the base of

Fig. 28. *Cat. 17.* Chest microscope with garnet lens.

Fig. 29. *Cat. 17.* Garnet lens.

References:
1. (_____) 'Sir David Brewster's Microscope', *Journal of the Royal Microscopical Society*, 18 (1898), 123–124; F. A. B. Ward, *A Catalogue of European Scientific Instruments in the Department of Medieval and Later Antiquities of the British Museum* (London, 1981), 138. For Brewster's family, see item 30 in this exhibition.
2. G. L'E. Turner, 'The Rise and Fall of the Jewel Microscope 1824–1837', *Microscopy*, 31 (1968), 92; Ward, *op cit.*, (1), 138.
3. D[avid] B[rewster], 'Microscope' *Encyclopaedia Britannica* 22 vols. 8th edition, (Edinburgh, 1856), 765–766; R. H. Nuttall and A. Frank, 'Makers of Jewel Lenses in Scotland in the Early Nineteenth Century', *Annals of Science*, 30 (1973), 407–416; Peter Hill is discussed in T. N. Clarke, *Scottish Scientific Instruments from the the Frank Collection in the Royal Scottish Museum (forthcoming)*.
4. A description of how this type of microscope dismantles is given in 'Old Portable Microscope by Dollond', *Journal of the Royal Microscopical Society*, 26 (1906), 713–715.
5. A. N. Disney *et al.* (eds.), *Origin and Development of the Microscope* ... (London, 1928), 194–195.

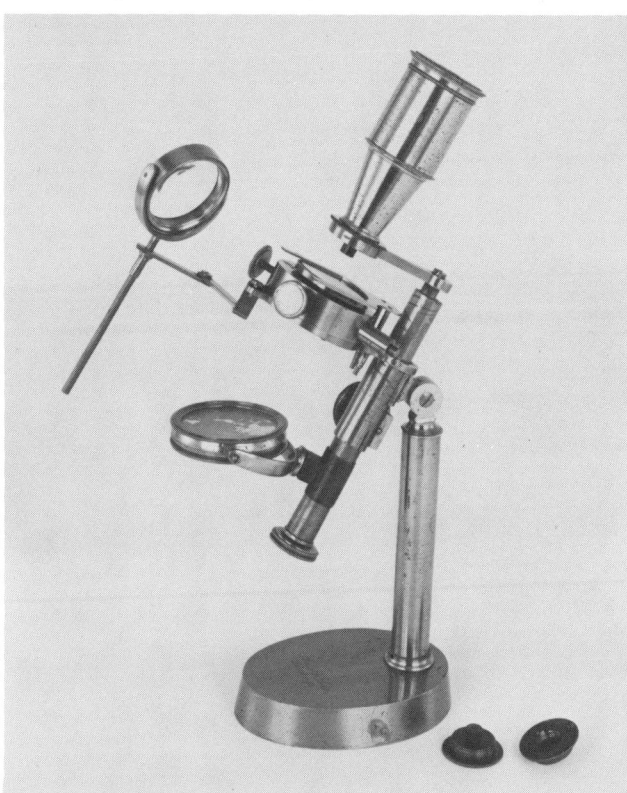

Fig. 30. *Cat. 18.* Jewel lens microscope by Adie & Son.

the pillar, above the trunion joint. One eye piece is present, and six objectives; one of the objectives, focal length about $\frac{1}{5}$ of an inch, is made of garnet. A brass plaque inside the box reads: 'Sir David Brewster/MADE ALL HIS EXPERIMENTS WITH/THIS MICROSCOPE TILL 1838./IT CONTAINS/THE FIRST GARNET LENS'. Another similar brass plaque on the outside lid reads 'PRESENTED/IN MEMORIAM/OF/SIR DAVID BREWSTER'.
Size of box: 325 × 139 × 127 mm.
Microscope body: 220 × 45 mm diameter.
Microscope pillar: 250 × 13 × 13 mm.

Provenance: Presented to the British Museum by Mrs. Brewster Macpherson, Brewster's daugher-in-law, in 1897[1] (Inv. BM 97 12-23.1). Presumably the wording of the engraved brass plaques may be attributed to her. If the instrument does indeed contain, as it is claimed 'the first garnet lens'[2], then this lens was probably made by Peter Hill of Edinburgh[3]. Apart from this single unusual feature, the instrument is very much anonymous, and a rather poor example of its type[4], its importance resting with its provenance. It has been suggested that it may possibly have been made by Dollond[5].

18. Jewel lens microscope by Adie & Son

Simple and compound jewel lens microscope, in brass. Rack and pinion coarse focusing mechanism moves the entire limb with body tube through stage attachment; fine focus is by micrometer screw at the base of the limb. Mechanical stage, concave mirror, bull's eye stage condensor. Stage is attached to trunion at top of pillar, which is set into oval eccentric base engraved 'Adie & Son/Edinburgh'. With two simple lenses, marked focal lengths $\frac{3}{4}$ and $\frac{3}{10}$ inches; two Coddington lenses, focal lengths $\frac{1}{10}$ and $\frac{1}{20}$ inches; one Wollaston doublet, focal length $\frac{1}{4}$ inch; and a garnet lens, focal length $\frac{1}{100}$ inch.
Size of instrument: 130 × 60 × 270 mm.
Size of box: 185 × 125 × 85 mm.

Provenance: From the Frank Collection of Scientific Instruments[1]. Acquired by the Royal Scottish Museum in 1979 (Inv. RSM TY 1979.47). Previously acquired by Mr. Frank from an auction in Paisley, about 1970. Exhibited at Kelvingrove Museum and Art Gallery, Glasgow, in the 'Tools of Science' exhibition, 1973[2].

The inscription on the microscope dates its sale to after 1835[3] at which time it is extremely unlikely that jewel lenses would have been in production. It has been suggested that the garnet lens, which would have been an expensive item in its time, came from unsold residual stock[4]. There are points of similarity between this item and other Adie microscopes in public collections in the United Kingdom, for example, an incomplete simple and compound instrument in the Wellcome Collection, at the Science Museum, London[5]. A number of similar features are found on a microscope retailed by Adie & Son, used by the amateur scientist and authoress Mary Somerville, a contemporary of Brewster, currently on loan to the Museum of the History of Science, Oxford.

References:
1. R. H. Nuttall, *Microscopes from the Frank Collection 1800–1860* (Glasgow, 1979), 34–35, No. 16; Norman Tebble (ed.), *The Royal Scottish Museum Triennial Report 1977–78–79* (Edinburgh, 1979), 49.
2. R. H. Nuttall, *The Arthur Frank Loan Collection: Early Scientific Instruments* (Glasgow, 1973), 56. No. 713.
3. For the dates and addresses of the Adie firm, see D. J. Bryden, *Scottish Scientific Instrument-Makers 1600–1900* (Edinburgh, 1972), 43; and T. N. Clarke *Scottish Scientific Instruments from the Frank Collection at the Royal Scottish Museum (forthcoming)* for a further discussion of the Adie business.
4. R. H. Nuttall and A. Frank, 'Makers of Jewel Lenses in Scotland in the Early Nineteenth Century', *Annals of Science, 30* (1973), 414; Nuttall, *op. cit.* (1), 35.
5. Accession No. A56356, signed 'Adie Edinburgh', purchased at Stevens Auction Rooms, London, 24 October 1920, lot 132.

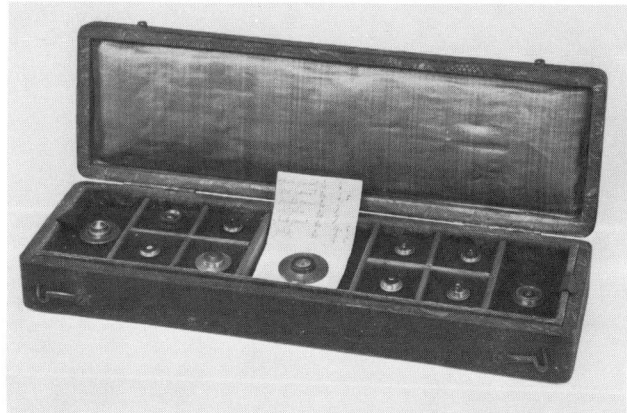

Fig. 31. *Cat. 19.* Set of jewel lenses.

19. Set of jewel lenses, attributed to Alexander Adie, Edinburgh

Morocco-covered case containing, among other optical pieces, five jewel lenses in brass mounts painted black, unsigned. Small manuscript note reads:

'pale garnet $\frac{1}{20}$ 4.0
Red garnet $\frac{1}{20}$ 4.5
Purple garnet $\frac{1}{30}$ 7.0
Sapphire $\frac{1}{30}$ 7.0
Pale garnet $\frac{1}{50}$ 9.5
glass $\frac{1}{30}$ 4.5'

Box lined inside with green silk and green felt.
Size (of lenses): 10 mm diameter thread.
Box: 360 × 105 × 60 mm.

Provenance: From the Frank Collection of Scottish Scientific Instruments, and presented to the Royal Scottish Museum in 1981 (Inv. RSM TY 1981.36). Discovered by Arthur Frank amongst material from 18 Forrest Road, Edinburgh[1], the opticians' shop previously run by J. R. and G. Hutchison, sons of George Hutchison who had been an apprentice to Adie and Wedderburn, successors to Alexander and John Adie[2].

Exhibited at Kelvingrove Museum and Art Gallery, Glasgow in the 'Tools of Science' exhibition, 1973[3]. There are other jewel lenses and jewel lens microscopes in the following UK public collections: British Museum, London; Science Museum, London; Wellcome Museum at the Science Museum, London; Whipple Museum of the History of Science, Cambridge; Museum of the History of Science, Oxford[4].

References:
1. R. H. Nuttall and A. Frank, 'Jewel lenses—a Historical Curiosity', *New Scientist, 53* (1972), 92–93; R. H. Nuttall and A. Frank, 'Makers of Jewel Lenses in Scotland in the Early Nineteenth Century', *Annals of Science, 30* (1973), 406–416.
2. J. R. Hutchison, 'Notable Opticians of Old Reekie', in *Programme of the 9th Annual Conference of the Scottish Association of Optical Practitioners* (Edinburgh, 1939), 17; D. J. Bryden, *Scottish Scientific Instrument Makers 1600–1900* (Edinburgh, 1972), 27, n. 132.
3. R. H. Nuttall, *The Arthur Frank Loan Collection: Early Scientific Instruments* (Glasgow, 1973) 56. No. 172.
4. Listed in G. L'E. Turner, 'The Rise and Fall of the Jewel Microscope 1824–1837', *Microscopy, 31* (1968), 85; also see item 18 in this exhibition.

Fig. 32. *Cat. 20.* Mechanical spinner.

20. Mechanical Spinner

Wooden spinner, consisting of tripod base with pillar (does not unscrew) with 2 pulley wheels at either side of the top, feeding looped string from circular turning handle to brass wheel set into end of adjustable wooden arm. Brass spindle through brass wheel has 2 holes (possibly for attachments, now missing) and arm has screw hole at other end. No maker's name, but label on one portion of foot. This reads: 'This was used by *Brewster*/(for) experimenting on/Prismatic Colours/(...)n Burnet 1856'.
Size (diameter of turn wheel): 240 mm.
Instrument: 315 × 310 × 660 mm.

Fig. 33. *Cat. 20.* Label on spinner base.

Provenance: Bought by Hertford Museum from a local antique dealer in 1968[1]. Earlier history unknown.

This piece resembles an item identified as a spinner which demonstrates 'the difference between static and dynamic equilibrium, where the position is governed by both gravitational and centrifugal forces'[2]. It appears from the evidence of the label on the foot that this particular spinner was modified, but it is now incomplete. 'Burnet' on the label may have been the Scots-born painter and engraver John Burnet, who left Edinburgh for London in 1806 to work with Sir David Wilkie[3]. Although there is no positive evidence that Burnet knew Brewster, they had various interests in common, notably improvements of mechanical processes of engraving[4], an interest in colour theory[5], and both were Fellows of the Royal Society of London. Brewster certainly knew other artists who were interested in colour: one, the minister of Duddingston parish outside Edinburgh, John Thomson, visited him at Balavil in 1833[6]; another, who visited Thomson in 1818 and may have met Brewster and discussed colour theory with him, was J. M. W. Turner[7].

References:
1. Correspondence from Hertford Museum, April 1981.
2. G. L'E. Turner, 'Van Marum's Scientific Instruments in Teyler's Museum: Descriptive Catalogue' in E. Lefebvre and J. G. de Bruijn, *Martinus van Marum Life and Work* 4 vols. (Leyden, 1973), IV, 163. No. 48, (Inventory No. 60) Fig. 51.
3. C[osmo] M[onkhouse], 'John Burnet' in Leslie Stephen (ed.), *Dictionary of National Biography* 53 vols (London, 1886) VII, 406–407.
4. *Ibid*; A. D. C. Simpson, 'Brewster's Society of Arts and the Pantograph Dispute', *(forthcoming)*.
5. J. Burnet, *Practical Hints on Colour* (London, 1827), 6, quotes Sir Isaac Newton.
6. W. Baird, *John Thomson of Duddingston: Pastor and Painter* (Edinburgh, 1895), 102–103; W. Baird, *Annals of Portobello* (Edinburgh, 1898), 484–486.
7. Discussed in John Gage, *Colour in Turner: Poetry and Truth* (London, 1969), 122–126.

21. Adjustable slit

Two rectangular brass plates with tapering inner edges can be moved in parallel by two L-shaped hinges. One of these plates is mounted on a trunion joint attached to a circular wooden base, through which an ivory knob connects. By turning the knob, the instrument can be tilted. A brass rod, connected to the top hinge raises and lowers the second

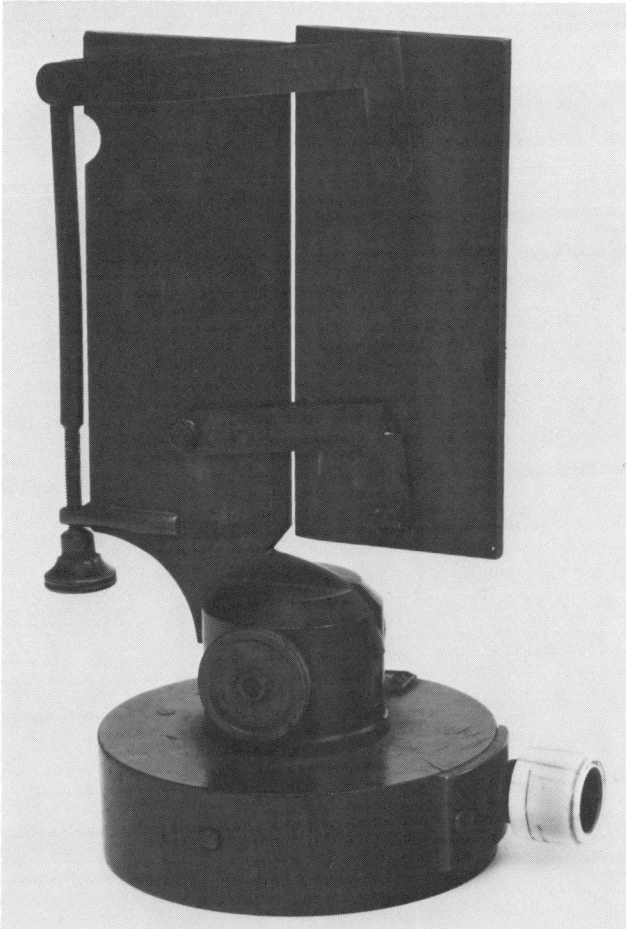

Fig. 34. *Cat. 21.* Adjustable slit.

plate, thus altering the space between the two plates. The upper hinge is teeped to one side with '1' and '2' screws. The ivory knob and wooden base are both stamped 'ROBISON'.
Size: 100 × 78 × 157 mm.

Provenance: Presented to the Royal Scottish Museum by the Natural Philosophy Department, University of Edinburgh in 1973[1] (Inv. RSM TY CU-92).

The 'Apparatus Register' of the University of Edinburgh Natural Philosophy Class Museum lists a number of items given by Sir David Brewster[2]: although there is no date of donation, the items were presumably presented during his years as Principal, from 1859 until his death in 1868. He is known to have used the Natural Philosophy Department's equipment and laboratories during this time[3]. This piece may have been used by Brewster in his experiments on solar spectral analysis[4] and bears a superficial resemblance to a polarimeter described and illustrated in one of his papers[5]. The stamp 'ROBISON' indicates that this piece of apparatus once belonged to John Robison, professor of natural philosophy at the University of Edinburgh from 1773 to his death in 1805, and teacher and friend of the young Brewster[6].

References:
1. Norman Tebble and R. Oddy (eds.), *The Royal Scottish Museum Triennial Report 1971–73* (Edinburgh, 1973), 60.
2. One manuscript 'Register' is held by the Department of Technology in the Royal Scottish Museum, and was compiled in 1905 from earlier registers. The four entries that relate to Brewster were described in 1905 as follows:

   'CU-84 Decayed glass (3 specimens) from Sir D Brewster (now 4 specimens, perhaps 1 broken).

CU-91 Ground glass plates (ordinary ground glass) from Sir D Brewster.
CU-92 3 slits—one mounted (and adjustable) from Sir D Brewster.
CU-228 Apparatus made by Sir D Brewster & presented to the Museum. Appears to be apparatus used by him in studying polarisation of light of the sky.'

Unfortunately, only the adjustable slit from CU-92 was extant when the Natural Philosophy Collection came to the Royal Scottish Museum.

3. C. G. Knott, *Life and Scientific Work of Peter Guthrie Tait* (Cambridge, 1911), 69.
4. For a discussion of Brewster's theory of the spectrum in an early nineteenth century context, see Paul D. Sherman, *Colour Vision in the Nineteenth Century: the Young-Helmholtz-Maxwell Theory* (Bristol, 1981); and for his work on spectroscopy, see M. A. Sutton, 'Sir John Herschel and the Development of Spectroscopy in Britain', *British Journal of the History of Science*, 7 (1974), 42–60.
5. David Brewster, 'On the Compensations of Polarised Light, with the Description of a Polarimeter for Measuring Degrees of Polarisation', *Transactions of the Irish Academy*, 19 (1843), 377–392.
6. M. M. Gordon, *The Home Life of Sir David Brewster* (Edinburgh, 1869), 36.

Fig. 35. *Cat. 22.* Early daguerreotype camera, possibly by Thomas Davidson.

22. Early daguerreotype camera, possibly by Thomas Davidson

Camera, in brass, mounted on trunion joint above pillar which screws below into flat circular base. Cylindrical body, with tube holding plano-convex and plano-concave combination lens with built-in stop, focused by a milled screw below the lens tube. Inside the body tube is a 60 mm square groundglass focusing screen. The single-sided brass plate-holder containing a prepared copper plate is pushed into a groove in front of the ground glass screen. The push-fit brass cap is replaced behind this, making the camera body light-tight. By turning the knob under the body tube, the plate-holder, plate, and groundglass screen are pivoted 180°, this action causing a projection inside the tube to flip open the plate-holder and expose the plate. After exposure by turning the knob back, the plate-holder closes and a new one, with an unexposed plate can immediately be inserted. This instrument is unsigned.
Size: 250 × 250 mm high.
Body tube diameter: 105 mm.
Lens aperture approx. 25 mm; focal length approx. 165 mm: f/6.6.
Plate size: 60 × 60 mm.

Provenance: RSM Register 1925 'April. Antique Camera, in brass, body $4\frac{1}{4}$ ins diameter, $6\frac{1}{2}$ ins long on pillar stand, with one single brass dark slide. Presented by Mr. Jas. Adams, Comrie Park, Dunfermline'. There is no correspondence relating to the accession of this object. (Inv. RSM TY 1925.16).

Although Helmut Gernsheim claimed to have been the first to identify this as a camera in July 1952[1], the 1925 Register entry above shows that it had always been recognized as such. Gernsheim's claim for Thomas Davidson as the undoubted maker of this piece cannot be fully substantiated. Davidson undoubtedly made photographic apparatus for Sir David Brewster[2], and for Hill and Adamson[3], besides selling 'Daguerreotype and Calotype Apparatus, with the latest improvements'[4] to a more general public.

Born into a Northumbrian labouring family in 1798[5], after an apprenticeship to a weaver, Davidson was happiest constructing telescopes, culminating in a successful demonstration to the British Association in Cambridge in 1833[6]. In 1836 he moved to Edinburgh, being employed as an instrument maker firstly with John Davis[7] then Robert Bryson[8], and then on his own account. With the arrival of the news of Daguerre's photographic discovery in 1839, Davidson enthusiastically took to practising photography[9], contributing papers on the art to various journals[10] for many years. However, he appears to have been a poor businessman, and after about fifteen years in Edinburgh, he returned to poverty and obscurity in Northumberland, dying in 1878[11].

References:
1. Helmut and Alison Gernsheim, *History of Photography* (London, 1969), 152.
2. David Brewster, 'A Brief Account of the Camera Obscura, and other Apparatus used in making Daguerreotype Drawings', *Report of the British Association for the Advancement of Science . . . 1840* (London, 1841), Part II, 9; A. D. Morrison-Low, 'Brewster and Scientific Instruments', in this volume; for Davidson, see D. J. Bryden *Scottish Scientific Instrument Makers 1600–1900* (Edinburgh, 1972), 46.
3. Referred to in John Nicol, 'Reminiscences of Thomas Davidson, a Weaver lad', *British Journal of Photography*, 26 (1879), 390–391, 399–401. A camera lens, used by Hill and Adamson, and possibly made by Davidson, now in the Science Museum, London (Inv. Sc. M. 1931–239), is described in D. B. Thomas, *The Science Museum Photography Collection* (London, 1969), 27, item 167. It was presented in turn by Mrs. D. O. Hill to R. T. Cochrane (24 November 1883), then by him to J. Craig Annan, (Christmas 1911), and subsequently by him to the Royal Photographic Society. It was exhibited by the Scottish Arts Council in 1970: Katherine Michaelson, *David Octavius Hill 1802–1870 and Robert Adamson 1821–1848* (Edinburgh, 1970), 78, item 295.
4. *Edinburgh Post Office Directory* (1844–45), 427, Advertisement.
5. Nicol, *op. cit.* (4), 390.
6. Thomas Davison, *(sic)* 'On a Reflecting Telescope', *Report of the British Association for the Advancement of Science . . . 1833*, (London, 1834), Part II, 420.
7. Bryden, *op. cit.* (2), 46.
8. T. N. Clarke, *Scottish Scientific Instruments from the Frank Collection at the Royal Scottish Museum (forthcoming)*.
9. He exhibited a 'Photogenic Camera' at an exhibition organised by the Society of Arts, December 1839–January 1840: *Catalogue of the Exhibition of Arts, Manufacturers, and Practical Science, in The Assembly Rooms, George Street* (Edinburgh, 1840), 25. It won the Scottish Society of Arts prize for *1840–41: Transactions of the Royal Scottish Society of Arts 2*. Appendix 1.

Among the papers he wrote on photography were several which were read at meetings of the Royal Scottish Society of

Arts: 'Description of the process of Daguerreotype, with Specimens and Diagrams Demonstrating the Action of Light in that Process, both in respect to Landscape and Miniature Portraits', *Proceedings of the Scottish Society of Arts*, 1 (1840), 76–77; *Transactions of the Royal Scottish Society of Arts*, 2 (1844), 21–25; *Edinburgh New Philosophical Journal*, 30 (1841), 178–182. 'Description and Drawing of a simple but Important Improvement in the Camera Obscura, in taking portraits and other objects' and 'Description and Diagrams of a method of Taking Views by Reflection in the Daguerreotype, or in the Camera Obscura', *Proceedings of the Scottish Society of Arts*, 1 (1841), 79–80; these three papers won Davidson the Society's Silver Medal for session 1840–41. On 12 June 1843 'Description and diagram of a Compound Achromatic Camera', by Mr. Thomas Davidson, FRSSA, MGSE and Hon Member Lit. & Phil Soc. of St. Andrews, Optician Edinburgh, was read to the Society in which 'a number of very beautiful specimens of Daguerreotype and of Calotype taken by Major Playfair and by Mr. Adamson by means of this camera, both portrait and landscape, were exhibited'. *Proceedings of the Royal Scottish Society of Arts*, 2 (1844), 53. This further links Davidson to the group of photographic experimentalists which surrounded Brewster at St. Andrews in the 1840s. Davidson also wrote a pamphlet about photography: *The Art of Daguerreotyping with the Improvements of the Process and Camera* (Edinburgh, 1841).
10. Including several in the *British Journal of Photography*.
11. Nicol, *op. cit.* (4), 401.

Lacock estate, but there is a possibility that between 1839 and 1841 they may have been supplied by the London optician Andrew Ross[3]. Like those similar to it in the Science Museum, this camera has an inspection hole, probably used for focusing[4], rather than checking the exposure of photogenic drawings[5]: the light entering would ruin the picture.

References:
1. See item 10 for provenance of this collection.
2. Inventory number Sc. M. 1937-346, described in D. B. Thomas, *The Science Museum Photography Collection* (London, 1969), 8. No. 11.
3. J. P. Ward, 'The Fox Talbot Collection at the Science Museum', *History of Photography*, 1 (1977), 277–278.
4. *Ibid.*
5. Helmut and Alison Gernsheim, *History of Photography* (London, 1969), 83.

Fig. 37. *Cat. 24.* Printing frame for calotypes.

24. Printing frame used by W. H. F. Talbot
Wooden printing frame in unvarnished white pine, used for exposing prints in the calotype process. Outside frame has fitted glass held in place with adhesive paper. Internal piece of wood is covered with padded white calico, nailed in place; there are two wooden handles to lift it out of the frame. A third block of wood holds this in place with wooden screws clamping it through a bridge piece attached to the outside frame.
Size: 350 × 300 × 205 mm.

Provenance: RSM Register 1936 'Original heavy printing frame (calotype) 14" × 12". Presented by Miss M. Talbot, Lacock, Abbey, Chippenham Wilts'. (Inv. RSM TY 1936.22)[1].

This printing frame is similar to one in the Science Museum[2], and like the preceding camera, item 23, would have been fairly easy to construct. It was used for printing calotypes[3] at Lacock, the photographic process patented in England by W. H. F. Talbot in 1839[4].

References:
1. See item 10 for the provenance of this collection.
2. Inventory no. Sc. M. 1937-354.
3. For a description of the calotype process, and of items from the Science Museum's Fox Talbot Collection, see D. B. Thomas, *The Science Museum Photography Collection* (London, 1969), 57.
4. Patent No. 8194, 1839.

Fig. 36. *Cat. 23.* Calotype camera used by W. H. F. Talbot.

23. Calotype camera used by W. H. F. Talbot.
Wooden box camera, lacking back or plate holders, in stained white pine. With metal peephole cover in top left corner, and metal lens sleeve screwed into centre of camera body. The lens unscrews from the push-fit focus holder, and is unmarked.
Size: 245 × 220 × 250 mm.

Provenance: RSM Register 1936 'Early Calotype camera with lens c1840 $9\frac{5}{8}'' \times 9\frac{5}{8}'' \times 8''$ over lens. Presented by Miss M. Talbot, Lacock Abbey, Chippenham, Wilts'. (Inv. RSM TY 1936.21)[1].

This camera is similar in construction to one in the Science Museum, also from Lacock Abbey: it would have been cheap and easy to make[2]. It has been suggested that Talbot's cameras were made by local carpenters on the

25. Box Iron used by W. H. F. Talbot
Small box iron, in iron, with turned wooden handle. The

Fig. 38. *Cat. 25.* Box iron, used in the calotype process.

back plate lifts up to allow a heated triangle of lead to be placed within; both lead and the inside of the back plate are stamped '4½'. The outside of the back plate is also stamped 'W. BULLOCK & CO'.
Size: 113 × 82 × 140 mm.

Provenance: RSM Register 1936 'Iron, fitted with bolt 4½" long 3¼" heel used by Fox Talbot in photographic process. Presented Miss M. Talbot, Lacock Abbey, Chippenham Wilts'[1]. (Inv. RSM TY 1936.96).

This box iron is similar to one in the Science Museum, also from Lacock Abbey[2]. Talbot used an ordinary domestic iron of the period to apply wax to calotype negatives to make them more transparent, and also to help dry the positives[3]. William S. Bullock, iron and steel merchant, is recorded at Great Charles Street, Birmingham, from 1817 until 1850[4].

References:
1. See item 10 for the provenance of this collection.
2. Inventory No. Sc.M. 1937-353.
3. D. B. Thomas, *The Science Museum Photography Collection* (London, 1969), 57. Item 380 refers to the use of box irons in the calotype process. Talbot's own notebooks record that he used the domestic iron for a variety of experiments: 10 August 1839 'Waterloo paper placed on blotting paper, and ironed on the back *with wax*; then silvered'; 20 January 1840 'Sheets of paper that have been ironed with a hot iron, are attracted very strongly by a table, &c. and repel each other — this electrical effect lasts a long time, and *might* have revealed that science to the ancients, if they had observed it'; 30 March 1840 '... Wash paper with white of egg, then iron it, the heat will solidify it'; 21 April 1840 'To coat paper with silver. Put a coating of bromide silver on it, of the consistence *(sic)* of curd; press a hot iron on it, which reduces the silver: then burnish. The iron does not decompose the bromide unless *wet*.'; 13 September 1842 '... The paper (A) is better made by drying it with an iron, for this makes the texture closer and less bibulous — When pictures are made on bibulous paper, they are apt to darken in the interior, or at any rate to come out with a coarser grain'. Fox Talbot Papers, Science Museum Library Archive Collection, Science Museum London. I am grateful to J. P. Ward for these references.
4. Communication from City of Birmingham Reference Library, giving dates from Birmingham trade directories.

26. Calotype Album
Purple leather-bound volume of some 300 pages, marked 'CALOTYPES' on the spine in gold tooled letters, filled with photographs, including some of the earliest calotypes taken in Scotland. These range in size and quality, and include some negatives; among the photographers can be identified Dr. John Adamson, Robert Adamson, Robert Adamson with D. O. Hill, and Thomas Rodger. The album appears to have been compiled by Dr. John Adamson late in his life, and is annotated by him.
Size: 275 × 355 × 60 mm.

Fig. 39. *Cat. 26.* Page 31 from calotype album, showing five calotypes of St. Andrews: St. Mary's College, the Cathedral, Blackfriars Chapel, the City Wall and the Episcopal Church. Lower two prints signed "RA".

Provenance: RSM Register 1942 'Calotype Photographs, among earliest in Scotland in two albums
1. 14" × 11" × 2¼" Calotype album.
2. 11½" × 10" × 1" Scrapbook [not exhibited in 1980-81].
Presented by Mr. W. P. H. Tulloch, Caledonia Club c/o East India and Sports Club, St. James's Square, London SW1 per Dr. R. O. Adamson, 15 Grosvenor Terrace, Glasgow W2.' (Inv. RSM TY 1942.1.1).

R. O. Adamson MD was a son of Dr. John Adamson and nephew of Robert Adamson[1]; W. P. H. Tulloch was John Adamson's grandson. Both this album (1942.1.1.) and the scrapbook (1942.1.2.) include historically significant material which is comparatively unknown[2]. Dr. John Adamson has attracted, as yet, little attention from historians of photography[3], but his brother Robert, particularly in conjunction with the artist D. O. Hill, has received some of the acclaim he deserves[4]. Brewster's correspondence[5] with Talbot had kept the Scotsman in touch with the invention of positive/negative photography, and because Talbot's 1839 patent did not apply to Scotland Brewster was able (with Talbot's approval) to encourage a number of his St. Andrews friends to take up the new art[6]. It would appear that Sir David tried his hand at photography, and it is known that he borrowed a Davidson camera from St. Andrews University[7].

The album contains what Dr. John Adamson claimed to be the first photographic portrait taken in Scotland[8]: a copy of this was exhibited at an Arts Council Exhibition in 1970[9]. The album comprises: experiments in the chemical 'fixing' of calotype images; photographs sent to Brewster by Talbot; successfully fixed calotypes of St. Andrews, including scenes to which Robert Adamson would return with his partner D. O. Hill[10]; and then a larger number of portraits, the majority of them taken by the wet collodion method, and possibly by John Adamson's pupil Thomas Rodger.

References:
1. In September 1950 Dr. R. O. Adamson wrote to the historian of photography Harold White: 'I was just 5 years old when my father died in 1870 and never realised till many years after the interest and value of my father's and uncle's work in photography which followed the advice of the late Mr. Fox Talbot, and strangely as it now seems to me, no one in my later life ever made much reference to it'. In a letter dated 5 April 1942, addressed to the Director of the Royal Scottish Museum, he wrote 'The two volumes presented to the Museum were first in my mother's possession, and then in my sister's—and finally in her son's, after my sister's decease years ago. This son was and has been a great deal abroad. All the Calotypes etc were, at his request, kept safely by me, for him. He really was and is the possessor and he was most anxious that the volumes if acceptable, should be donated to the Museum.' The nephew was W. P. A. Tulloch.
2. See for example, brief mention in Colin Ford and Roy Strong, *An Early Victorian Album: The Photographic Masterpieces (1843–1847) of David Octavius Hill and Robert Adamson* (New York, 1976), 15–16. For a more detailed discussion of these albums and the work of the Adamson brothers see A. D. Morrison-Low, 'Dr. John and Robert Adamson: An Early Partnership in Scottish Photography', *The Photographic Collector 4* (1983), 134–147.
3. Helmut and Alison Gernsheim, *The History of Photography* (London, 1969), 164–167; H. J. P. Arnold, *William Henry Fox Talbot: Pioneer of Photography and Man of Science* (London, 1977), 145.
4. Sara Stevenson, *Robert Adamson and David Octavius Hill: A catalogue of the collection held by the Scottish National Portrait Gallery* (Edinburgh, 1981): this contains the most complete bibliography of secondary literature on Hill and Adamson.
5. Only Brewster's side of this correspondence survives, at the Science Museum, London, in the Science Museum Library Archive Collection (Fox Talbot Papers), and at Lacock Abbey, Chippenham, Wiltshire.
6. A photograph album exists in the Jammes Collection, Paris, annotated, compiled and possibly photographed by Brewster: a microfilm copy of this is in Edinburgh City Library. A number of the prints are duplicates of examples in the first RSM album, and presumably these were taken by different members of the St. Andrews circle. The RSM albums contain some prints, the negatives of which have been initialled 'RA' 'JA' and 'TP'.
7. Discussed under item 22 of this exhibition.
8. Adamson, however, altered the annotated date in the album from '1841' to an unrealistically early '1840': Helmut Gernsheim, 'Some Common Misconceptions about Early Photography in Scotland', *Functional Photography*, 4 (1952), 12–13.
9. Katherine Michaelson, *David Octavius Hill 1802–1870 and Robert Adamson 1821–1848* (Edinburgh, 1970), 33–34, Items 80 and 81, Plates 12 and 13.
10. Graham Smith, 'Hill and Adamson: St. Andrews, Burnside and the Rock and Spindle', *The Print Collector's Newsletter 10* (1979) 45–48; 'Hill and Adamson at St. Andrews: the Fishergate Calotypes', *ibid.*, *12* (1981), 33–37.

27. Microphotographs by J. B. Dancer of Manchester
A selection of microphotographs by John Benjamin Dancer from his commercially available range together with an early example of his first known slide, which depicted the

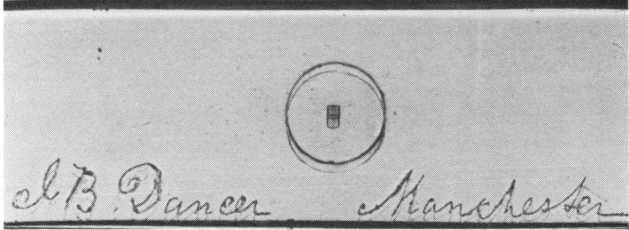
Fig. 40. *Cat. 27a.* Early microphotograph by J. B. Dancer.

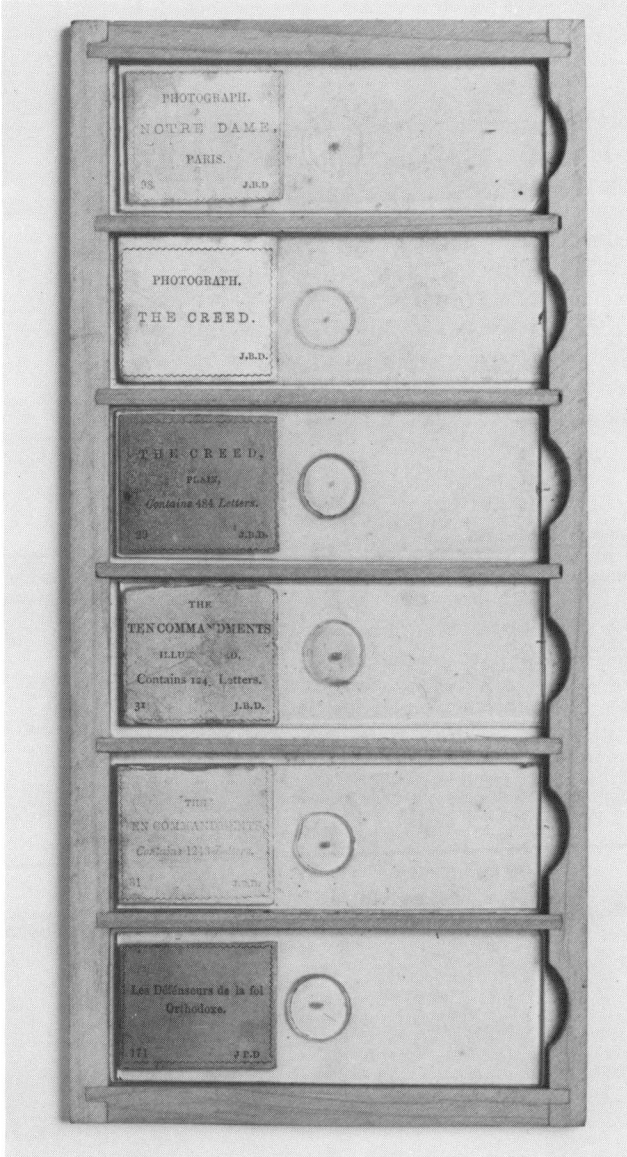
Fig. 41. *Cat. 27.* Five commercially produced microphotographs by J. B. Dancer.

memorial tablet of William Sturgeon, the electrician and inventor, at Kirby Lonsdale Church.
a. 'Sturgeon's Memorial Tablet', engraved on glass 'J. B. Dancer Manchester'
b. 'The Creed'
c. 'The Creed 29'
d. 'The Ten Commandments 31'
e. 'The Ten Commandments Illuminated 31'
f. 'Notre Dame 98'
g. 'Les Defennseurs de la foi Orthodoxe 171'
Each of these, except the first, has a yellow label glued to the left of the cover slip; in the lower right corner of each label are the printed initials 'JBD'.
Size: 75 × 25 × 2 mm each.

Provenance: Purchased by the Royal Scottish Museum in

1980; the example showing Sturgeon's memorial tablet was previously acquired in Chester. (Inv. RSM TY 1980.63 1–7).

Dancer made the first recorded microphotographs, that is, photographs which are reduced so that they have to be viewed through a microscope, a few months after Daguerre announced his photographic process in 1839[1]; but these were only practical to produce in very limited numbers. However, with the advent of the wet collodion process in 1852 Dancer was able to use this to produce the first commercial microphotographs[2], and these were enthusiastically regarded as curiosities during his lifetime. The so-called 'diamond signed' microphotographs appear to be early experiments which Dancer presented to friends[3]. Sir David Brewster suggested a number of possible uses for the invention[4] including the proposal that 'Microscopic copies of despatches and valuable papers and plans might be transmitted by post, and secrets might be placed in spaces not larger than a full stop or a blot of ink'[5]. Brewster took several examples of Dancer's work on his Continental tour in 1857[6] where they were seen by the French photographer Prudent René Patrice Dagron, who later used the method during the siege of Paris in 1870 to send messages by carrier pigeons across the enemy lines[7].

References:
1. G. W. W. Stevens, 'Microphotography since 1829', *The Photographic Journal*, 90 (1950), 149–156.
2. The priority dispute over microphotographs between George Shadbolt and J. B. Dancer is discussed by H. Milligan in 'New Light on J. B. Dancer', *Memoirs and Proceedings of the Manchester Literary and Philosophical Society*, 115 (1973), 1–9.
3. H. Milligan, 'On the Dating of Early Microphotographs', *Bulletin of the European Society for the History of Photography*, 1 (1980), 1–2. The 'Sturgeon's Memorial Tablet' microphotograph is also discussed in L. L. Ardern, 'John Benjamin Dancer: Instrument maker, Optician and the Originator of Microphotography', *Library Association, North Western Group:* Occasional papers No. 2 (London, 1960).
4. D[avid] B[rewster], 'Micrometers' in *Encyclopaedia Britannica 14* 22 vols. 8th edition, (Edinburgh, 1856) 745.
5. D[avid] B[rewster] 'Microscope' in *Encyclopaedia Britannica 14* 22 vols. 8th edition, (Edinburgh, 1856) 802.
6. M. M. Gordon, *The Home Life of Sir David Brewster* (Edinburgh, 1869), 280.
7. A. McLeod, 'John Benjamin Dancer: Originator of Microphotography', *British Journal of Photography*, 120 (1973), 138–141. A. T. Gill has produced '*J. B. Dancer—An Annotated Bibliography*', (London, 1972) which was reproduced in M. Hallett (ed.), 'John Benjamin Dancer 1812-87. Selected Documents and Essays', (printed 1979); the same compilation also contained A. L. E. Barron's *List of J. B. Dancer Microphotographs, with Notes* (London, 1974–75). Both Gill and Barron originally had their listings produced for the Royal Photographic Society.

28. Lenticular stereoscope by Duboscq

Collapsable Brewster-type stereoscope in wood and black cloth, with internal wire frame, in which a stereo pair of photographs mounted on a card is viewed through lenses at a distance of about 90 mm, the average distance between the eyes. Stamped with 'DS' monogram, surrounded by 'BREVETÉ/S.G. du G.T'
Size: 110 × 155 × 100 mm.

Provenance: Lent, and subsequently presented by Thomas H. Court to the Science Museum, London (Inv. Sc. M. 1918–166).

The history of Brewster's second best-selling invention[1]

Fig. 42. *Cat. 28.* Lenticular stereoscope by Duboscq.

is as difficult to unravel as that of his first[2]. Brewster was, however, always careful to insist that he invented the lenticular form of the instrument, and not the stereoscope as such[3]. Despite attempts to interest various British scientific instrument makers, including George Lowdon[4] of Dundee, it was not taken up as a commercial enterprise until Brewster's visit to Paris in 1850[5], when he showed the Parisian firm of Duboscq et Soleil an example of Lowdon's work[6]. Duboscq et Soleil specialised in optical apparatus, particularly polarising equipment[7]. At the Great Exhibition in 1851 an example of Brewster's stereoscope was displayed[8] which 'attracted the attention of the Queen, and M. Duboscq manufactured one, which Sir David Brewster presented to Her Majesty in the name of the maker'[9]. The DS monogram stands for Duboscq-Soleil, while 'BREVETÉ/S.G. du GT' (sans garantie du gouvernement) is the legal requirement for a French patentee[10].

References:
1. David Brewster, *The Stereoscope: its History, Theory and Construction, with its Application to the Fine and Useful Arts and to Education etc.* (London, 1856).
2. Arthur T. Gill, 'Early Stereoscopes', *The Photographic Journal*, 109 (1969), 546–559, 606–614, 641–651.
3. M. M. Gordon, *The Home Life of Sir David Brewster* (Edinburgh, 1869), 345–346.
4. A. H. Millar, *James Bowman Lindsay and other Pioneers of Invention* (Edinburgh, 1925), 86.
5. Gordon, *op. cit.* (3), 347–348.
6. No Lowdon stereoscope has been traced, although microscopes (e.g. Inv. RSM TY 1980. 238 and A56557 in the Wellcome Museum) and telescopes (eg. RSM TY 1980.268 and 1981.42) and a camera (eg. RSM TY 1981.41) are known to have been either made or retailed by Lowdon.
7. The firm was founded by Soleil in 1819, and he later took his son-in-law Jules Duboscq into partnership; by the end of the century, the firm was run by Ph. Pellin, as it states on the front of his trade catalogue: *Instruments d'Optique et de Précision* (Paris, 1900).
8. *Great Exhibition of the Works of Industry of All Nations 1851. Official Descriptive and Illustrative Catalogue* 4 vols. (London, 1851) III, 1235, entry 1197.
9. Gordon, *op. cit.* (3), 348.
10. J. Johnson & J. H. Johnson, *The Patentee's Manual: A Treatise on the Law and Practice of Patents and Inventions* (6th edition, London, 1890), 437–440.

29. Lenticular stereoscope

Brewster-type lenticular stereoscope, walnut veneered. With simple semi-lenses, adjustable mirror and opaque ground glass screen.
Size: 170 × 190 × 150 mm.

Provenance: Presented by A. Johnston to the Royal Scottish Museum in 1952 (Inv. RSM TY 1952.25).

Fig. 43.   *Cat. 29.* Lenticular stereoscope.

Following its reception at the 1851 Great Exhibition, and its approval by Queen Victoria, Brewster's stereoscope became an overnight success in Britain, and was produced in enormous numbers and a variety of different forms[1]. Although both French and British manufacturers took out numerous patents to cover their improvements to the design[2], it is noticeable that Brewster himself did not do so.

References:
1. R. S. Clay, 'Presidential Address: The Stereoscope', *Transactions of the Optical Society*, 29 (1927–28), 149–166.
2. Arthur T. Gill, 'Early Stereoscopes', *The Photographic Journal*, 109 (1969), 546–559, 606–614, 641–651.

Fig. 44.   *Cat. 30.* Stereophotograph, possibly taken by Sir David Brewster.

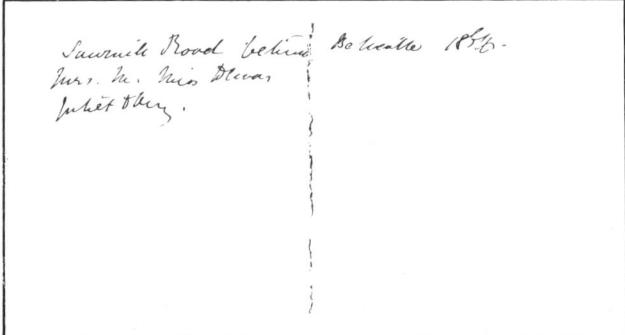

Fig. 45.   *Cat. 30.* Reverse of stereophotograph.

30.   Stereo Photograph, attributable to David Brewster
Stereo pair of wet collodion photographs, possibly taken by Sir David Brewster. The image depicts four figures in a wooded Highland landscape; the two parts are individually cut from a double print and mounted on yellow card, which has been folded down the centre. The reverse is inscribed in Brewster's handwriting: 'Sawmill Road behind Belville 1864. Mrs M. Miss Dewar Juliet & (? May).'
Size of card: 172 × 85 mm
Images: 73 × 69 mm each (approx).

Provenance: Lent by Mrs. M. Brewster-Macpherson; from Brewster's papers still held by the family[1].

This appears to be the only photograph of what must have been an extensive collection, to have survived the 1903 fire. It shows a part of the family estate, of Belleville or Balavil by Kingussie, which had come to Brewster through his first wife Juliet (neé Macpherson) and which by the 1860s was being administered by Brewster's son, who had changed his name to Brewster-Macpherson[2]. Brewster, with his second wife and young daughter, visited his son in Inverness-shire and his daughter in Aberdeenshire each summer[3] towards the end of his life: he had administered the Balavil estate himself on behalf of his sister-in-law, and his wife between 1833 and 1836[4]. The figures in the photograph appear to be friends or house-guests rather than family.

Brewster's interest in photography had begun with the earliest British work, and he took an active interest in its progress in Scotland; he was involved in both the Photographic Society of Scotland, and the Edinburgh Photographic Society[5]. His invention of the lenticular stereoscope[6] led to a new photographic craze, comparable only with the popularity of the kaleidoscope. Brewster suggested a binocular stereoscopic camera as early as 1847[7] but it was not for another two years that he described this[8]. Reviewing this latter article, anonymously, he stated that 'Mr Slater of Euston Square has already constructed several of these binocular cameras, which have been sent to America'[9]. None are known to have survived.

The first stereoscopic camera taking simultaneous pairs of photographs was produced in 1853 by the Manchester optician J. B. Dancer, and was inspired by Brewster's lenticular stereoscope[10]. There was some confusion about the patenting of this, and the only surviving example was destroyed in an air raid in 1940[11]; however, Dancer patented a later version, the first stereoscopic camera to be offered for sale to the general public, in 1856[12]. It is not known whether Brewster owned or used either form of Dancer's binocular camera.

References:
1. National Library of Scotland Temp. Dep. 1837; listed in National Register of Archives (Scotland) Survey No. 718.
2. Genealogy supplied by Mrs. M. Brewster Macpherson.
3. M. M. Gordon, *The Home Life of Sir David Brewster* (Edinburgh, 1869), 355.
4. *Ibid.*, 154–160.
5. The Photographic Society of Scotland was founded in 1856 while the Edinburgh Photographic Society was formed in 1862: for a discussion of these rival bodies and an explanation why the former foundered while the latter survived, see Sara Stevenson, 'David Octavius Hill without Robert Adamson: the Reputation of their Calotypes after 1848', *The Photographic Collector*, 2 (1981), No. 3, 14–24.
6. See item 28 in this exhibition.
7. [David Brewster], 'Photography', *North British Review*, 7 (1847), 502.
8. David Brewster, 'Account of a Binocular Camera, and of a Method of Obtaining Drawings of Full Length and Colossal Statues and of Living Bodies which can be Exhibited as Solids by the Stereoscope', *Transactions of the Royal Scottish Society of Arts*, 3 (1851), 259–264, but delivered to the Society Monday 9 April 1849 (communication 3242) *ibid.*, Proceedings of the Royal Scottish Society of Arts, Appendix Q, 217.

9. [David Brewster], 'Binocular Vision and the Stereoscope', *North British Review*, *17* (1852), 181.
10. J. B. Dancer, 'An Autobiographical Sketch', *Memoirs of the Manchester Literary and Philosophical Society*, *107* (1964–65), 115–142.
11. A. T. Gill, 'Early Stereoscopic Cameras' *Photographic Journal*, *115* (1976), 42–43, 114–115.
12. B. Coe, *Cameras: From Daguerreotypes to Instant Pictures* (London, 1978), 155–157.

Appendix 1(b)

Brewster's Contribution to the Study of the Lens of the Eye: An Experimental Foundation for Modern Biophysics.

GEORGE DUNCAN

In the early years of the nineteenth century, while continental war was raging, European eye anatomists were also locked in conflict over the nature of the substance within the lens capsule. It concerned the matter of fact and artifact and has been a recurring theme of conflict ever since. Leeuwenhoek and Sattig maintained that they could see fibrous structures in an extract of lens that had been prepared for examination, but in a long and elegant reply in Latin, Soemmerring argued that the fibres were simply an artifact of the method of preparation[1]. Brewster decided that the only way to resolve this dispute was to develop methods for investigating lens structure without having to recourse to a water or alcohol extract.

background, and, moreover, the bright image of the lens had a Maltese Cross appearance. (Fig. 46)[2]. These observations led Brewster to conclude that the lens was birefringent and that the ordered structures giving rise to the birefringence were probably radially oriented with respect to the optic axis of the lens. He therefore reasoned that these structures were the fibres proposed by the early anatomists and set out to prove this in a series of experiments of great simplicity and insight[3]. He cut a small section from a cod lens, carefully noting the region of the lens where the piece was removed and, maintaining the orientation of the section with respect to the optic axis of the lens, he mounted it in the centre of a brass ring. With the section mounted vertically so he could view the light from a candle passing through the optic axis of the specimen, he found that there were coloured images on the horizontal plane on either side of a bright central image (Fig. 47). Brewster realised that the images arose from light

Fig. 46. The Maltese-Cross appearance of the lens when viewed through crossed polaroids. Brewster considered that the inner structure arose from the radial orientation of the lens fibre cells and he noted that the outer, smaller structure had the opposite sign of birefringence from the inner. This structure probably corresponds to the tangentially-oriented layers of lens capsule. (From Brewster, 1816).

In 1816 Brewster had noted that, when viewed through a polarising microscope, the lens appeared bright on a dark

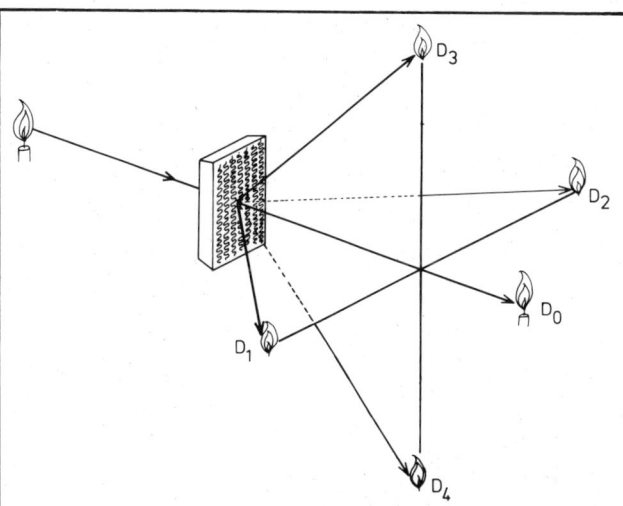

Fig. 47. Brewster's Optical Diffraction system. He dissected out small pieces of lens and placed them in an annular holder. On viewing a candle through the section, he observed a series of diffraction patterns (D_1, D_2, D_3 and D_4) and from the spacing between D_1 and D_2 calculated the fibre spacing within the lens. The value of 4.8 μm that he obtained is perfectly acceptable today. From the fact that he also observed vertical patterns (D_3 and D_4) he concluded that there were regular protrusions on the surface of the fibres.

diffracted from the regularly repeating structures in the lens tissue and he chose the clearest image, namely the red, for his calculations. He in fact also knew that Fraunhöfer had recently derived a formula that would enable him to calculate the spacing of his ordered units, but he preferred to derive the measurement with the aid of a clever model system. He had a fellow scientist, John Barton, make several sets of very close, accurate rulings on a steel bar and he viewed the reflected light from these. He found that the set of rulings that gave the same angular separation of the two red images as he had obtained when viewing the tissue were spaced 1/5000th of an inch apart. After correcting for the fact that he was comparing the images from light directly incident on the lens and obliquely incident on the steel, Brewster calculated the repeat distance of the fibres to be 4.8 μm. From the variation in spacing of the patterns obtained from different parts of the lens, Brewster suggested the model for lens fibre cell structure shown in Fig. 48a.

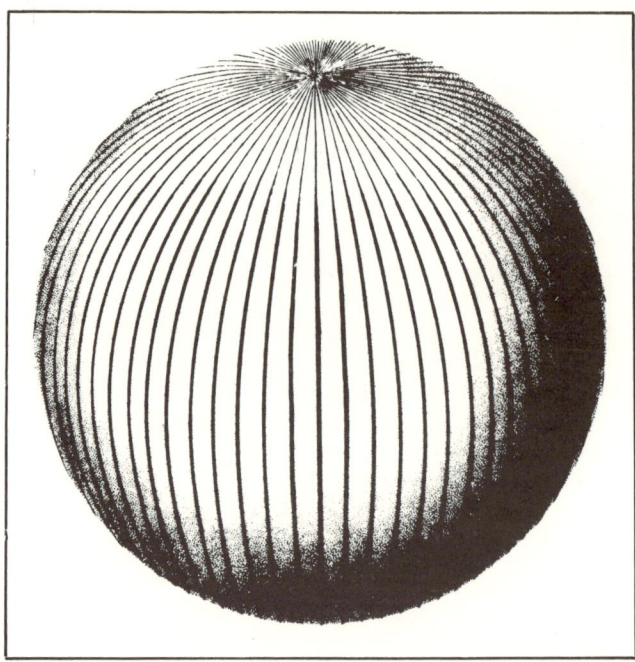

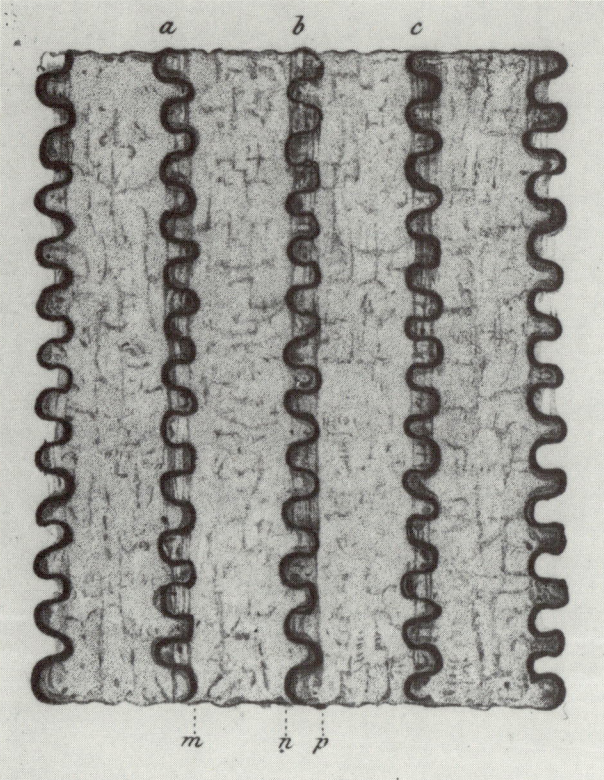

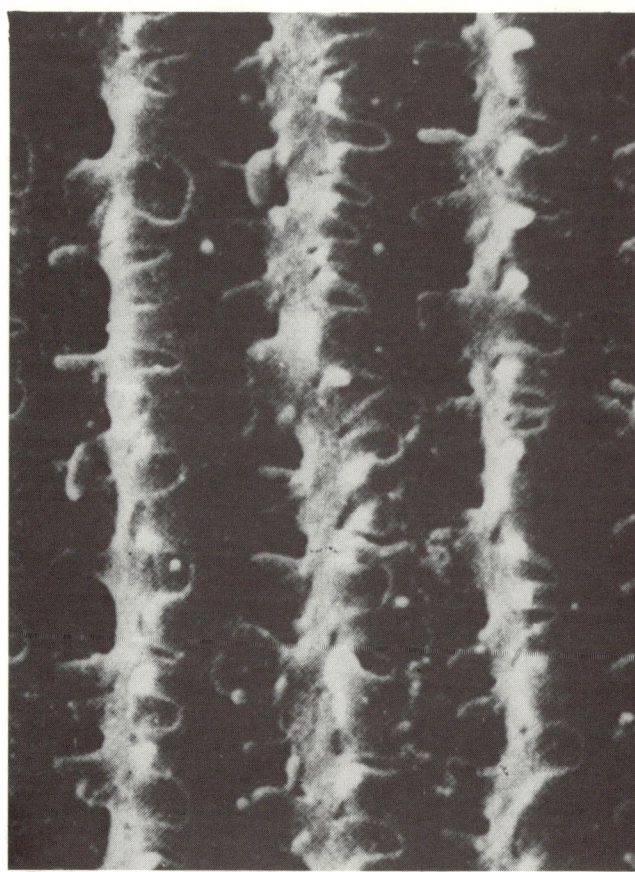

Fig. 48. (a) Brewster's model of the arrangement of fibre cells within the lens. He discovered that the fibre diameter varied throughout the lens and that the tips met in a series of suture lines on the optic axis of the lens. (From Brewster, 1833).

Brewster also noted that red diffraction patterns were obtained in the vertical plane when viewing the candle through the tissue and found that the angular spacing of these was approximately five times greater than that obtained in the horizontal plane. He therefore reasoned that they arose from regularly repeating structures at right angles to the long axis of the lens fibres. He examined a small piece of lens in the best microscope available (which he did not identify) and indeed observed invaginations on the surface of the fibres (Fig. 48b). He further noted that: 'The two fibres ab, bc are united by a series of teeth, exactly like those of a rack work, the projecting teeth of the one fibre entering into the hollows between the teeth of the adjacent one. The breadth of the teeth is such that five of them are nearly equal to nm + np, which is a measure of the breadth of a fibre'[4].

This is surely one of the best descriptions of the ball-and-socket joints between lens fibres, recently rediscovered by using scanning electron microscopic techniques on the lens

(b) Brewster's drawing of adjacent fibre cells with their 'ball and socket' interdigitations. This structure was not seen again by lens anatomists until after the advent of the scanning electron microscope. (From Brewster, 1833).

(c) Scanning electron microscope view of adjacent cortical fibre cells. (From Clark *et al.*, 1980).

(Fig. 48c)[5]. This wealth of data obtained from what are basically simple optical diffraction patterns, contrasts starkly with the lack of detail in laser diffraction patterns reported recently[6].

However Brewster's optical studies did not end there. He returned to the lens from which he had removed the small section and viewed the light from his candle after it had been reflected by the cut edge. He was delighted to find that the images were again coloured and also very distinct and made the following important observation: 'The perfect flatness of the surfaces of the concentric laminae, as indicated by their power of forming a distinct image by reflexion, shows that the fibres which compose them are flat and not cylindrical'[7]. He then made an imprint of the cut surface of the lens on to isinglass and to his surprise and delight, he found the same coloured images when he viewed the light reflected from the isinglass replica. He had therefore proved without doubt that the diffraction patterns were *a property of the order of the system and not of the substance of the system*.

In this elegant series of experiments Brewster was using a number of mathematical and physical techniques to determine the ultrastructural and physiological basis of a biological tissue and I know of no better example of a publication that may be truly said to lay foundations for modern biophysics. In summary, the advances he may be said to have made are as follows. He developed non-invasive techniques (in this case optical diffraction) to resolve a dispute of fact and artifact. He made quantitative measurements of the structure he was investigating. He applied modern mathematical techniques in his calculations (in this case from Fraunhöfer's Wave Theory, even although he was a devoted advocate of Newton's corpuscular theory). He checked these calculations by simulating his observations with a defined model system. He understood that the patterns he obtained were more a property of the order of the biological system than of the substance of the system, thus substantially prefiguring the philosophical and practical foundations on which modern X-ray crystallography is based. He also understood the practical and medical consequences of his observations and put forward suggestions concerning the nature and cure of cataract[8].

That Brewster has received scant recognition for laying this firm foundation would have surprised him not at all and is probably due to the 'aphakik'[9] view of science engendered by successive educational systems rather than any lack of lustre in the illumination of science emanating from our elders.

Notes and References
1. S. T. von Soemmerring, *Icones Oculi Humani* (Frankfurt, 1804) quoted in D. Brewster, 'On the Anatomical and Optical Structure of the Crystalline Lenses of Animals, particularly that of the Cod'. *Philosophical Transactions of the Royal Society of London*, 123 (1833), 323–332.
2. David Brewster, 'On the Structure of the Crystalline Lens in Fishes and Quadrupeds as Ascertained by its Action on Polarised light', *Philosophical Transactions of the Royal Society of London*, 106 (1816), 311–317.
3. Brewster, *op. cit.* (1).
4. David Brewster, 'On the Development and Extinction of Regularly Doubly Refracting Structures in the Crystalline Lenses of Animals after Death', *Philosophical Transactions of the Royal Society of London*, 127 (1837), 253–258.
5. J. I. Clark, L. Mengel, A. Bagg and G. B. Benedek, 'Cortical Opacity, Calcium Concentration and Fibre Membrane Structure in the Calf Lens', *Experimental Eye Research*, 31 (1980), 399–410.
6. G. Duncan and T. J. C. Jacob, 'The Lens as a Physico-chemical System' in H. Davison (ed.) *The Eye* (2nd edition, London, *in press*), I.
7. Brewster, *op. cit.* (1).
8. David Brewster, 'On the cause and cure of cataract', *Transactions of the Royal Society of Edinburgh*, 24 (1865), 11–14.
9. '*Aphakia, aphacia* (medical). The condition of the eye when the lens has been removed': T. C. Collocot (ed.), *Chambers Dictionary of Science and Technology* (Edinburgh, 1971), 60.

Appendix II(a)

Papers of Sir David Brewster in the National Library of Scotland

PATRICK CADELL

As a result of the fire at Balavil in 1903, there is no longer extant a large collection of Brewster's own papers, but because of his wide ranging interests and extraordinary energy, he was in touch with a large number of his contemporaries, and many of his letters to them have survived as well as a number of copies of their letters to him. The National Library of Scotland has a large quantity of Brewster material that has survived in this way. The following is a list of it. The arrangement is chronological. Where a recipient only is given it can be assumed that the letter is from Brewster; where only a writer appears then Brewster is the recipient. Where no folio reference is given it can be assumed that the manuscript has an index.

Letters to Dr. Robert Anderson (1800–12), Adv. Ms 22.4.13, ff. 212–221.
Letters to Robert Lundie, Minister of Kelso, (1808), Ms 9847, ff. 275, 278v; (1809, 1829), Ms 9848, ff. 1, 188.
Correspondence with the Very Rev. John Lee concerning *The Edinburgh Encyclopaedia*, (1809–10), Ms 3432. ff. 199–295; (1811–12, 1814), Ms 3433, ff. 11, 37, 82, 153, 161, 258–261, 264–267; (1827), Ms 3436, f. 247; (1837), Ms 3341, f. 310; (1840), Ms 3443, ff. 16, 27, 36; (1846), Ms 3445, ff. 177, 194, 227, 237; (1849–50), Ms 3446, ff. 167, 200; (1851), Ms 3447, f. 22, (n.d.), Ms 3449, ff. 16–17. Key to writers in the *Edinburgh Encyclopaedia* (1833), Ms 3456.
Letters to various correspondents (1810–65) including Thomas Campbell, Dr. Thomas Clark, Thomas Constable, and Professor James Pillans, Ms 1808, ff. 28–67.
Letters from Blackwoods (1814, 1825–29). Acc. 5643, A 1–3, B 6.
Letters to George Combe WS (1815), Ms 7201, ff. 159–162; (1821), Ms 7206, f. 14.
Letters of Archibald Constable and Co., (1817, 1819), Ms 790; (1802–22), Ms 791; (1823–24), Ms 792.
Letters to Blackwoods, (1817), Ms 4002; (1819), Ms 4004; (1821), Ms 4006; (1824), Ms 4012; (1825), Ms 4014; (1826), Ms 4016; (1832), Ms 4032; (1838), Ms 4046; (1856), Ms 4115; (1862), Ms 4168; (1867), Ms 4218; (n.d.), Ms 4714, ff. 88–102.
Letter of Sir Walter Scott (c1817), Ms 997, f. 120.

Letters to Robert Stevenson (1818, 1821), Ms 19988, ff. 23, 30.
Letter of Leonard Horner to Dr. John Marcet (1819) concerning Brewster, Ms 9818, f. 82.
Letters to Thomas Carlyle (1819–20), Ms 1764, ff. 149, 177, 193; (1824, 1828), Ms 1765, ff. 18, 84.
Letters to James Skene (1819, 1824–29, 1831, n.d.), Ms 3813, ff. 75–101.
Letters to Professor T. S. Traill (1819–20), 19338, ff. 162, 195–199, 204; (1821–24), 19339, ff. 32–225 *(passim)*; (1827), 19341, f. 19; (1833), 19345, f. 42.
Correspondence concerning the Royal Scottish Society of Arts (from 1819), Acc. 4534 *passim*.
Letter to Sir Walter Scott (1820), Ms 1583, f. 93.
Letter of Professor T. S. Traill (1820), Ms. 19390, f. 87.
Letters to various correspondents including William Chambers (1820, 1849), Ms 581, nos 490, 520; (1832, 1854), Ms 583, nos 754 A–B; (1854), Ms 584, no. 990.
Correspondence with Sir Walter Scott (1821), Ms 867, f. 138; Ms 1750 f. 291; (1823), Ms 3897, f. 181; (1825), Ms 3901, ff. 95, 211.
Letters to John Gibson Lockhart (1825, 1833, n.d.), Ms 925, nos 10–12.
Memorandum and letters to 2nd Earl of Minto (1826), Ms 11811, f. 43; (1826–7); Ms 12226, ff. 87, 91–101; (1827); Ms 11914, f. 200; (1832); Ms 13409, ff. 83, 107; (1836); Ms 11916, f. 52; Ms 12049, f. 91; (1837), Ms 12058, f. 28.
Letter of Sir Patrick Walker to Donald Horne concerning Brewster (1827), Ms 14299, f. 85.
Letter to Alexis Bouvard (1828), Acc 8227.
Letter to Blackwoods (1829), Acc 5308.
Letter to A. Scott, Kelso (1832), Adv. Ms 7.1.19, f. 8.
Letters to Robert Paul, the Banker, (1832), Ms 5139, f. 161; (1850), Ms 5140, f. 92.
Letters to W. H. Lizars (1832–59), Ms 1831, ff. 25–45, 48–52.
Letters to various correspondents (1837, 1842, 1854, 1860, 1862, 1867), Ms 9752, ff. 25–35.
Letter to Viscount Melville (1838), Ms 16, f. 66.
Letter to Lord Advocate Murray (1838), Ms 98, f. 12.
Letter to Professor James Pillans (1841), Ms 2522, f. 3.
Letters to Hugh Miller (1842, 1854), Ms 7527, nos 26, 34.
Letters to William Mure of Caldwell (1843), Ms 4948, ff. 191–196, 200; (1844–45), Ms 4949, ff. 24, 60, 71–75, 217.

Fig. 49. Sir David Brewster. Calotype by D. O. Hill and R. Adamson. *(Scottish National Portrait Gallery)*.

Letters to Hon. W. H. Percy (1845, 1851), Acc. 5387.
Letters to Professor J. S. Blackie (1845, 1851), Ms 2622. ff. 21, 221, 248; (1857), Ms 2624, f. 294; (1859), Ms 2625, f. 121; (1867), Ms 2628, f. 216.
Letter to (? Robert) Chambers (1848), Ms 5174, f. 4.
Letter to Dr. Moir (1850), Ms 7179, no. 42.
Letter of Mary Ward (1856), Ms 9940.
Letter of Prosper Mérimée to Edward Ellice (1856), concerning Brewster, Ms 15038, f. 10.
Letter to J. F. Campbell of Islay (1858), Adv. Ms 50.7.7, f. 21.
Letters to Isabella Bird and another correspondent (1858, 1863), Ms 7178, nos 22, 25.
Letters to Thomas Stevenson (1859), Ms 785, ff. 51–53.
Letter to Mr. Herrick (1861), Ms 3650, f. 190.
Letter to Robert Omond (1862), Ms 19335, f. 262.
Letter to John H. Balfour-Browne (1863), Ms 19609, f. 4.
Letters to W. E. Aytoun (1863, n.d.), Ms 4896, ff. 217, 243.
Letters to William Robertson (1867–68), Ms 3952, ff. 122–124, 136, 140, 145.
Letter to E. Magrath (n.d.), Ms 962, f. 25.
Letter (n.d.), Ms 9812, f. 80.

Appendix II(b)

Published Writings of Sir David Brewster: a Bibliography

A. D. MORRISON-LOW

This bibliography was compiled in the first instance from *The Royal Society of London Catalogue of Scientific Papers* 19 vols., (London, 1867–1925), Walter E. Houghton (ed.) *The Wellesley Index to Victorian Periodicals* 3 vols., (Toronto, 1966–1979), the *British Museum General Catalogue of Printed Books to 1955* (New York, 1964), and Mrs. Gordon's list of her father's miscellaneous writings, recorded both in the text of her biography and in the Appendix: M. M. Gordon *The Home Life of Sir David Brewster* (Edinburgh, 1869). It was then supplemented by items found in a search through publications edited by Brewster, and through other journals, periodicals and encyclopaedias published during Brewster's lifetime, and discovered in family volumes of Brewster's bound offprints. Although it is considerably longer than any previous listing of Brewster's published writings, it is not expected that this listing can be complete, so numerous and diverse were his publications.

Unlike the Royal Society listing, this one gives each paper or book an individual number for each different date or place of publication, instead of a single number for different reprintings. Each entry has the source of information for believing the work to be by Brewster, and where appropriate Royal Society numbers are given. A list of abbreviations used is given below.

Books in Print	Whitaker's *British Books in Print* (London, 1939–1981).
B. M. Cat.	*British Museum General Catalogue of Printed Books to 1955* (New York, 1964).
EUL	*Catalogue of Printed Books in Edinburgh University Library* (Edinburgh, 1918–1923).
Family volume	Bound volumes of offprints owned by Brewster's descendants.
Gordon	M. M. Gordon, *The Home Life of Sir David Brewster* (Edinburgh, 1869).
List	List of authors, given at the beginning of the *Edinburgh Encyclopaedia* and each edition of *Encyclopaedia Britannica*.
Morse	E. W. Morse, 'Natural Philosophy, Hypotheses and Impiety: Sir David Brewster confronts the Undulatory Theory of Light'. Unpublished Ph.D. thesis, University of California, Berkeley, 1972.
Nat. Un. Cat.	*National Union Catalogue of Printed Books Pre-1956* (New York, 1966).
NLS	*Catalogue 1 [pre 1968] of Printed Books in the National Library of Scotland* (Edinburgh, 1974).
Poole	William Frederick Poole, *An Index to Periodical Literature* (Boston, Mass., 1882), and C. Edmund Wall (ed.), *Cumulative Author Index for Poole's Index to Periodical Literature 1802–1906* (Ann Arbor, 1971).
R. S. C.	*The Royal Society of London Catalogue of Scientific Papers* (London, 1867–1925).
Signet Library	*Catalogue of the Printed Books in the Library of the Society of Writers to H. M. Signet in Scotland* (Edinburgh, 1871).
Strout	Alan Lang Strout, *A Bibliography of Articles in Blackwood's Magazine 1817–1825* (Lubbock, 1959).
Sutton	M. A. Sutton, 'Sir John Herschel and the Development of Spectroscopy in Britain', *British Journal for the History of Science*, 7 (1974), 42–60.
W.I, W.II, W.III	Walter E. Houghton (ed.), *The Wellesley Index to Victorian Periodicals* (Toronto, 1966–1979).
Whewell	Letter from David Brewster to William Whewell, 19th November 1838, in Whewell Mss., Trinity College, Cambridge, MS. a. 201[81].

1.
'Observations on Picot's New Astronomical Theory', *Edinburgh Magazine, or Literary Miscellany*, 16 (1800), 278–281. (Signed article).

2.
D. B., 'On the Effects of the French Revolution upon Science and Philosophy', *Edinburgh Magazine, or Literary Miscellany*, 16 (1800), 287–288.

3.
D. B., 'On the Causes which tend to Retard and Accelerate Physical Science, exemplified in the History of Optics', *Edinburgh Magazine, or Literary Miscellany*, 16 (1800), 377–382.

4.
D. B., 'A Vindication of Newton's Opinion concerning the Figure of the Earth from the Objections of Henry B. St. Pierre, author of the *Studies of Nature*', *Edinburgh Magazine, or Literary Miscellany*, 16 (1800), 462–466.

5–10.
D. B., 'Celestial Phenomena for February 1801', *Edinburgh Magazine, or Literary Miscellany, 17* (1801), 3–5; '... for March 1801; with an attempt to explain the red colour of the Moon in lunar eclipses', *ibid.*, 133–136; '... for April', *ibid.*, 175–176; '... for May', *ibid.*, 303–305; '... for June', *ibid.*, 337–338; '... for July', *ibid.*, 462–464.

11–16
D. B., 'Celestial Phenomena for August 1801', *Edinburgh Magazine, or Literary Miscellany, 18* (1801), 46–47; '... for September', *ibid.*, 131–133; '... for October', *ibid.*, 169–171; '... for November', *ibid.*, 244–246; '... for December', *ibid.*, 355–357; '... for January 1802', *ibid.*, 399–402.

17.
'A Discourse on the Discoveries and Character of Sir Isaac Newton; read at a Meeting of Gentlemen in Edinburgh, on the Anniversary of his Birth-day, January 5th 1802', *Edinburgh Magazine, or Literary Miscellany, 19* (1802), 19–23. (Signed article).

18–23.
D. B., 'Celestial Phenomena for February 1802', *Edinburgh Magazine, or Literary Miscellany, 19* (1802), 37–39; '... for March', *ibid.*, 86–88; '... for April', *ibid.*, 165–167; '... for May', *ibid.*, 285–287; '... for June', *ibid.*, 323–325; '... for July', *ibid.*, 443–445.

24.
D. B., 'A few Notices Concerning the Planet lately discovered at Palermo, in Sicily, by M. Piazzi', *Edinburgh Magazine, or Literary Miscellany, 19* (1802), 46–48.

25.
D. B., 'Notice Concerning another new Planet Discovered by Dr Olbers of Bremen, in March 1802', *Edinburgh Magazine, or Literary Miscellany, 19* (1802), 287–288.

26.
D. B., 'Further Notices Respecting the Two New Planets, with some Remarks tending to shew that they cannot both belong to the Planetary System', *Edinburgh Magazine, or Literary Miscellany, 19* (1802), 445–448.

27.
D. B., 'Lines on the Death of Miss A[gnes] S[cott]', *Edinburgh Magazine, or Literary Miscellany, 19* (1802), 454–455.

28–33.
D. B., 'Memoirs of the Progress of Manufactures, Science and the Fine Arts', *Edinburgh Magazine, or Literary Miscellany, 20* (1802), 41–45, 83–88, 214–215, 290–292, 366–372, 450–457.

34–39.
D. B., 'Celestial Phenomena for August 1802', *Edinburgh Magazine, or Literary Miscellany, 20* (1802), 45–47; '... for September', *ibid.*, 89–90; '... for October', *ibid.*, 177–179; '... for November', *ibid.*, 292–295; '... for December', *ibid.*, 372–374; '... for January 1803', *ibid.*, 457–458.

40.
'Remarks on the Different Systems of the Universe', *Edinburgh Magazine, or Literary Miscellany, 20* (1802), 103–108. (Signed article).

41.
'The Invitation, an Elegy', *Edinburgh Magazine, or Literary Miscellany, 20* (1802). (Signed article).

42.
'Lines Written on Leaving Forfar, August 31, 1802', *Edinburgh Magazine, or Literary Miscellany, 20* (1802), 301–302. (Signed article).

43.
'Lines on Visiting the Grave of a Young Lady', *Edinburgh Magazine, or Literary Miscellany, 20* (1802), 379. (Signed article).

44–49.
D. B., 'Memoirs of the Progress of Manufactures, Science and the Fine Arts', *Edinburgh Magazine, or Literary Miscellany, 21* (1803), 43–45, 124–131, 205–207, 288–291, 364–367, 446–450.

50–55.
D. B., 'Celestial Phenomena for February 1803', *Edinburgh Magazine, or Literary Miscellany, 21* (1803), 45–47; '... for March', *ibid.*, 131–134; '... for April', *ibid.*, 207–209; '... for May', *ibid.*, 291–292; '... for June', *ibid.*, 367–368; '... for July', *ibid.*, 450–452.

56.
'Song', *Edinburgh Magazine, or Literary Miscellany, 21* (1803), 51–52. (Signed article).

57–62.
D. B., 'Memoirs of the Progress of Manufactures, Science and the Fine Arts', *Edinburgh Magazine, or Literary Miscellany, 22* (1803), 49–51, 131–133, 214–216, 290–291, 372–373, 444–446.

63–68.
D. B., 'Celestial Phenomena for August 1803', *Edinburgh Magazine, or Literary Miscellany, 22* (1803), 51–52; '... for September', *ibid.*, 133–134; '... for October', *ibid.*, 216–217; '... for November', *ibid.*, 291–292; '... for December', *ibid.*, 373; '... for January 1804', *ibid.*, 446–448.

69.
[Anon.], 'Biographical Account of the Celebrated French Astronomer M. Bailly', *The Scots Magazine, 65* (1803), 511–514. (? Editorial contribution).

70–81.
D. B., 'Memoirs of the Progress of Manufactures, Science and the Fine Arts' *The Scots Magazine, 66* (1804), 29–31, 111–113, 202–204, 286–288, 363–366, 407–408, 510–512, 605–608, 678–680, 735–736, 814–816, 894–896.

82–93.
D. B., 'Celestial Phenomena for February 1804', *The Scots Magazine, 66* (1804), 31–32; '... for March', *ibid.*, 109–111; '... for April', *ibid.*, 204–206; '... for May', *ibid.*, 288–290; '... for June', *ibid.*, 366–367; '... for July', *ibid.*, 405–407; '... for August', *ibid.*, 509–510; '... for September', *ibid.*, 604–605; '... for October', *ibid.*, 677–678; '... for November', *ibid.*, 733–735; '... for December', *ibid.*, 812–814; '... for January 1805', *ibid.*, 892–894.

94.
[David Brewster: With a dedication by William Alexander Laurie, to whom this work sometimes been attributed],
The History of Free Masonry, drawn from Authentic Sources of Information; with an account of the Grand Lodge of Scotland, from its institution in 1736, to the present time, compiled from the records; and an appendix of original papers (Edinburgh, 1804).

95.
[Anon.], *The History of Free Masonry and the Grand Lodge of Scotland. With chapters on the Knight Templars, Knights of St. John, Mark Masonry and R. A. Degree.* (Edinburgh, 1859). (A revised and enlarged edition of 94).

96.
[Anon.], *Geschichte der Freimanrerbrüderschaft in Schottland nach Will. Alex. Laurie's history of Free Masonry and the Grand Lodge of Scotland frei bearbeiter von Br. Bibliothekar Dr. Mersdorf.* (Cassel, 1861). (A revised and enlarged edition of 94).
(B. M. Cat.).

97–108.
D. B., 'Celestial Phenomena for February 1805', *The Scots Magazine, 67* (1805), 45–47; '... for March', *ibid.*, 123–124; '... for April', *ibid.*, 205–206; '... for May', *ibid.*, 253–255; '... for June', *ibid.*, 365–366; '... for July', *ibid.*, 414–415; '... for August', *ibid.*,

493–495; '... for September', *ibid.*, 574–575; '... for October', *ibid.*, 654–655; '... for November', *ibid.*, 734–735; '... for December', *ibid.*, 812–813; '... for January 1806', *ibid.*, 892–894.

109–120.
D. B., 'Memoirs of the Progress of Manufactures, Science and the Fine Arts', *The Scots Magazine*, 67 (1805), 47–48, 124–126, 207–208, 255–256, 366–368, 415–416, 495–496, 575–576, 655–656, 735–736, 813–814, 894–896.

121–123. James Ferguson
Lectures on select subjects in Mechanics, Hydrostatics, Pneumatics and optics; with the use of the Globes; the art of dialling; and the calculation of the mean times of reward full Moons and eclipses. A new edition corrected and enlarged with notes and an appendix adapted to the present state of the arts and sciences by D. Brewster. (Edinburgh, 1805).
—Second edition, 2 vols. (Edinburgh, 1806).
—Third edition, with plates, 2 vols. (Edinburgh, 1823).
(B. M. Cat.)

124.
[Anon: 'A Calm Observer'] 'An Examination of the Letter addressed to Principal Hill, on the case of Mr. Leslie, in a letter to its Anonymous Author: With remarks on Mr. Stewart's postscript, and Mr. Playfair's pamphlet'; in *Tracts Historical and Philosophical*, 2 (Edinburgh, 1806).
(B. M. Cat., Morse, Gordon.)

125.
'Remarks on Achromatic Eyepieces', *Journal of Natural Philosophy, Chemistry and the Arts: by W. Nicholson*, 14 (1806), 388–389.
(R. S. C. 1).

126–137.
D. B., 'Celestial Phenomena for February 1806', *The Scots Magazine*, 68 (1806), 7–8; '... for March', *ibid.*, 85–86; '... for April', *ibid.*, 166–167; '... for May', *ibid.*, 246–247; '... for June', *ibid.*, 327–328; '... for July', *ibid.*, 405–407; '... for August', *ibid.*, 494–495; '... for September', *ibid.*, 575–576; '... for October', *ibid.*, 654–655; '... for November', *ibid.*, 773–774; '... for December', *ibid.*, 814–815; '... for January 1807', *ibid.*, 894–895.

138–149.
D. B., 'Memoirs of the Progress of Manufactures, Chemistry, Science and the Fine Arts', *The Scots Magazine*, 68 (1806), 41–42, 86–87, 167–168, 247–248, 328, 407–408, 495–496, 576, 655–656, 774–775, 815–816, 895–896.

150.
D. B., 'Query respecting the Phenomena of Loch Ness', *The Scots Magazine*, 68 (1806), 405.

151.
D. B., 'Winter: A Dirge', *The Scots Magazine*, 68 (1806), 856.

152.
'Description of a New Astrometer for finding the rising and setting of the Stars and Planets, and their positions in the Heavens', *Journal of Natural Philosophy, Chemistry and the Arts; by W. Nicholson*, 16 (1807), 320–324.
(R. S. C. 2).

153.
'Description of a circular Mother-of-Pearl Micrometer', *The Philosophical Magazine ... edited by Alex. Tilloch*, 29 (1807), 48–52.
(R. S. C. 3).

154–164.
D. B., 'Celestial Phenomena for February 1807', *The Scots Magazine*, 69 (1807), 5–6; '... for March', *ibid.*, 115–116; '... for April', *ibid.*, 204–205; '... for May', *ibid.*, 253–254; [no entry for June] '... for July', *ibid.*, 406–408; '... for August', *ibid.*, 485–486; '... for September', *ibid.*, 566–567; '... for October', *ibid.*, 647–648; '... for November', *ibid.*, 726–727; '... for December', *ibid.*, 805–806; '... for January 1808', *ibid.*, 886–887.

165–175.
D. B., 'Memoirs of Manufactures, Chemistry, Science, and the Fine Arts', *The Scots Magazine*, 69 (1807), 6–7, 116–117, 205–206, 254–255, [362 by J. M.] 408, 486–487, 567, 687, 727–728, 806–808, 887–888.

176.
'Remarks on M. Burckhardt's Contrivance for shortening Reflecting Telescopes; with a New Method of making Refracting Telescopes with a Tube only one-third of the Focal Length of the Object-glass', *The Philosophical Magazine ... edited by Alex. Tilloch*, 33 (1809), 290–292.
(R. S. C. 4).

177.
'On the Fibres used in Micrometers; with an Account of a Method of Removing the Error Arising from the Inflection of Light, by Employing Hollow Fibres of Glass', *The Philosophical Magazine ... edited by Alex. Tilloch*, 33 (1809), 383–384.
(R. S. C. 5).

178.
[Anon.], 'Hydrodynamics', *Encyclopaedia Britannica*, 20 vols., 4th edition (Edinburgh, 1810), X, 697–792.
(List).

179.
[Anon.], '[Free] Masonry', *Encyclopaedia Britannica*, 20 vols., 4th edition (Edinburgh, 1810), XII, 638–668.
(List).

180.
[Anon.], 'Mathematics (history)', *Encyclopaedia Britannica*, 20 vols., 4th edition (Edinburgh, 1810), XIII, 1–14.
(List).

181.
[Anon.], 'Mechanics', *Encyclopaedia Britannica*, 20 vols., 4th edition (Edinburgh, 1810), XIII, 47–136.
(List).

182.
[Anon.], 'Optics', *Encyclopaedia Britannica*, 20 vols., 4th edition (Edinburgh, 1810), XV, 171–295. Original article by Jones, revised by Robison and subsequently by Brewster.
(List).

183. James Ferguson.
Astronomy explained upon Sir Isaac Newton's principles. With notes and supplementary chapters by D. Brewster, 2 vols., with a quarto volume of plates (Edinburgh, 1811).
(B. M. Cat.)

184–185.
'Demonstration of the fundamental Property of the Lever', *Transactions of the Royal Society of Edinburgh*, 6 (1812), 397–404; *Journal of Natural Philosophy, Chemistry, and the Arts; by W. Nicholson*, 30 (1812), 280–285.
(R. S. C. 6).

186.
'On some Properties of Light', *Philosophical Transactions of the Royal Society of London*, 103 (1813), 101–109.
(R. S. C. 7).

187.
A Treatise on New Philosophical Instruments for Various Purposes in the Arts and Sciences. With Experiments on light and colours. (London and Edinburgh, 1813).
(B. M. Cat.)

188.
'On the Affections of Light transmitted through Crystallised Bodies', *Philosophical Transactions of the Royal Society of London*, 104 (1814), 187–218.

189.
'On the Polarisation of Light by Oblique Transmission through all Bodies, whether Crystallised or Uncrystallised', *Philosophical Transactions of the Royal Society of London*, 104 (1814), 219–230.
(R. S. C. 9).

190–194.
'On the New Properties of Light exhibited in the Optical Phenomena of Mother-of-Pearl and Other Bodies to which the Superficial Structure of that Substance can be Communicated', *Philosophical Transactions of the Royal Society of London*, 104 (1814), 397–418; *Giornale di Fisica, Chemica e Storia Naturale; da L. Brugnatelli*, 8 (1815), 253–262; *Journal de Physique, de Chimie et de l'Histoire Naturelle; par J. C. de Lamétherie et Ducrotay de Blainville*, 81 (1815), 181–188; *Annals of Philosophy, or Magazine of Chemistry, Mineralogy, Mechanics, and the Arts; by T. Thomson*, 3 (1814), 190–196; *Blackwood's Edinburgh Magazine*, 2 (1817), 33–35, 140–142.
(R. S. C. 10, Strout).

195.
'Results of some Recent Experiments on the Properties impressed upon Light by the Action of Glass raised to different Temperatures, and cooled under different Circumstances', *Philosophical Transactions of the Royal Society of London*, 104 (1814), 436–439.
(R. S. C. 11).

196.
'On the Optical Properties of Sulphuret of Carbon, Carbonate of Barytes and Nitrate of Potash, with Inferences respecting the Structure of Doubly Refracting Crystals', *Transactions of the Royal Society of Edinburgh*, 7 (1815), 285–302.
(R. S. C. 12).

197.
'On a New Species of Coloured Fringes produced by the Reflection of Light between Two Plates of Glass of Equal Thickness', *Transactions of the Royal Society of Edinburgh*, 7 (1815), 435–444.
(R. S. C. 13).

198.
'Expériences sur la lumière', *Nouveau Bulletin des Sciences de la Société Philomathique de Paris*, (1815), 44–46.
(R. S. C. 14).

199.
'Additional Observations on the Optical Properties and Structure of Heated Glass and Unannealed Glass Drops', *Philosophical Transactions of the Royal Society of London*, 105 (1815), 1–8.
(R. S. C. 15).

200.
'Experiments on the Depolarisation of Light as exhibited by various Mineral, Animal, and Vegetable Bodies, with a Reference of the Phenomena to the General Principles of Polarisation', *Philosophical Transactions of the Royal Society of London*, 105 (1815), 29–53.
(R. S. C. 16).

201.
'On the Effects of Simple Pressure in producing that Species of Crystallisation which forms Two oppositely Polarised Images, and exhibits the Complementary Colours by Polarised Light', *Philosophical Transactions of the Royal Society of London*, 105 (1815), 60–64.
(R. S. C. 17).

202–203.
'On the Laws which regulate the Polarisation of Light by Reflexion from Transparent Bodies', *Philosophical Transactions of the Royal Society of London*, 105 (1815), 125–159; *Journal für Chemie und Physik; von J. S. C. Schweigger*, 17 (1816), 135–153.
(R. S. C. 18).

204.
'On the Multiplication of Images, and the Colours which Accompany Them in Some Specimens of Calcareous Spar', *Philosophical Transactions of the Royal Society of London*, 105 (1815), 270–292.
(R. S. C. 19).

205.
'On New Properties of Heat as Exhibited in its Propagation along Glass Plates', *Philosophical Transactions of the Royal Society of London*, 106 (1816), 46–114.
(R. S. C. 20).

206–207.
'On the Communications of the Structure of Doubly-Refracting Crystals to Glass, Muriate of Soda, Fluor Spar, and other substances by Mechanical Compression and Dilatation', *Philosophical Transactions of the Royal Society of London*, 106 (1816), 157–178; *Journal de Physique, de Chimie, et de l'Histoire Naturelle; par J. C. de Lamétherie et Ducrotay de Blainville*, 83 (1816), 80–85, 213–228, 309–330, 389–414.
(R. S. C. 21).

208–209.
'On the Structure of the Crystalline Lens in Fishes and Quadrupeds, as ascertained by its action on Polarised Light', *Philosophical Transactions of the Royal Society of London*, 106 (1816), 311–317; *Journal de Physique, de Chimie, et de l'Histoire Naturelle; par J. C. de Lamétherie et Ducrotay de Blainville*, 84 (1817), 379–383.
(R. S. C. 22).

210–211.
'On the Effects Produced in Astronomical and Trigonometrical Observations etc., by the Descent of the Fluid which Lubricates the Cornea', *The Journal of Science and the Arts*, 2 (1817), 127–131; *Annales de Chimie, ou Recueil de Mémoires concernant la Chimie et les Arts qui en dépendent*, new series, 4 (1817), 24–33.
(R. S. C. 23).

212.
'On the Decomposition of Light by Simple Reflexion', *The Journal of Science and the Arts*, 2 (1817), 211.
(R. S. C. 24).

213–215.
'On the Connection between the Primitive Forms of Crystals, and the Number of their Axes of Double Refraction', *Memoirs of the Wernerian Natural History Society*, 3 (1817–20), 50–74; *Journal de Physique; de Chimie et de l'Histoire Naturelle; par J. C. de Lamétherie et Ducrotay de Blainville*, 89 (1819), 36–50; *Annalen der Physik; von L. W. Gilbert*, 69 (1821), 1–29.
(R. S. C. 25).

216.
[Anon.], 'Hydrodynamics', *Encyclopaedia Britannica*, 20 vols., 5th edition (Edinburgh, 1817), X, 697–792.
(List).

217.
[Anon.], 'Mathematics (history)', *Encyclopaedia Britannica*, 20 vols., 5th edition (Edinburgh, 1817), XIII, 1–14.
(List).

218.
[Anon.], 'Mechanics', *Encyclopaedia Britannica*, 20 vols., 5th edition (Edinburgh, 1817), XIII, 47–136.
(List).

219.
[Anon.], 'Optics', *Encyclopaedia Britannica*, 20 vols., 5th edition (Edinburgh, 1817), XV, 171–295.
(List).

220.
[Anon.], 'Telescope', *Encyclopaedia Britannica*, 20 vols., 5th edition (Edinburgh, 1817), XX, 236–272.
(List).

221.
'Dr. Brewster's Patent Kaleidoscope', *Blackwood's Edinburgh Magazine, 3* (1818), 121–123.
(Strout).

222.
'History of Dr. Brewster's Kaleidoscope', *Blackwood's Edinburgh Magazine, 3* (1818), 331–337.
(Strout).

223.
'Sur le mouvement perpétuel', *Annales de Chimie, ou Recueil de Mémoires concernant la Chimie et les Arts qui en dépendent*, new series, 9, (1818), 219–220.
(R. S. C. 26).

224.
'Description du Kaleidoscope', *Bibliothèque Universelle des Sciences, Belles-Lettres, et Arts faisant suite à la Bibliothèque Britannique rédigée à Genève. Partie des Sciences*, first series, 8 (1818), 155–160.
(R. S. C. 27).

225.
'On the Action of Transparent Bodies upon the differently Coloured Rays of Light', *Transactions of the Royal Society of Edinburgh, 8* (1818), 1–23.
(R. S. C. 28).

226.
'Description of a New Darkening Glass for Solar Observations, which has also the Property for Polarising the whole of the Transmitted Light', *Transactions of the Royal Society of Edinburgh, 8* (1818), 25–30.
(R. S. C. 29).

227.
'On the Optical Properties of Muriate of Soda, Fluate of Lime, and the Diamond, as exhibited in their action upon Polarised Light', *Transactions of the Royal Society of Edinburgh, 8* (1818), 157–164.
(R. S. C. 30).

228.
'On a new Optical and Mineralogical Property of Calcareous Spar', *Transactions of the Royal Society of Edinburgh, 8* (1818). 165–170.
(R. S. C. 31).

229.
'On the Effects of Compression and Dilatation in altering the Polarising Structure of Doubly Refracting Crystals', *Transactions of the Royal Society of Edinburgh, 8* (1818), 281–286.
(R. S. C. 32).

230.
'On the Laws which Regulate the Distribution of the Polarising Force in Plates, Tubes, and Cylinders of Glass that have received the Polarising Structure', *Transactions of the Royal Society of Edinburgh, 8* (1818), 353–372.
(R. S. C. 33).

231.
'On the Laws of Polarisation and Double Refraction in Regularly Crystallised Bodies', *Philosophical Transactions of the Royal Society of London, 108* (1818), 199–272.
(R. S. C. 34).

232.
'Description of a Method of making Doubly Refracting Prisms perfectly Achromatic', *Annals of Philosophy, or Magazine of Chemistry, Mineralogy, Mechanics, and the Arts; by T. Thomson, 11* (1818), 175–177.
(R. S. C. 35).

233–234.
'On the Difference between the Optical Properties of Arragonite and Calcareous Spar', *The Journal of Science and the Arts, 4* (1818), 112–114; *Annales de Chimie, ou Recueil de Mémoires concernant la Chimie et les Arts qui en dépendent, 6* (1817), 104–106.
(R. S. C. 36).

235.
'Optical Structure of Ice', *The Journal of Science and the Arts, 4* (1818), 155.
(R. S. C. 37).

236.
[Anon.], *Description and Method of using the Patent Kaleidoscope invented by Dr. Brewster*. (London, 1818).
(B. M. Cat.).

237.
D. B., Review of Mackenzie's 'System of the Weather of the British Islands', *Blackwood's Edinburgh Magazine, 4* (1818), 84–87.
(Strout).

238.
D. B., 'Account of the Expedition to the North Pole', *Blackwood's Edinburgh Magazine, 4* (1818), 95–98.
(Strout).

239.
D. B., 'Account of Capt. Kater's New Method of Measuring the Pendulum', *Blackwood's Edinburgh Magazine, 4* (1818), 182–187.
(Strout).

240.
D. B., Analysis of Barrow's 'Voyages to the Arctic Regions', *Blackwood's Edinburgh Magazine, 4* (1818), 187–193.
(Strout).

241–243.
'On a new Optical and Mineralogical Structure, exhibited in Certain Specimens of Apophyllite and Other Minerals', *Edinburgh Philosophical Journal, 1* (1819), 1–8; *Transactions of the Royal Society of Edinburgh, 9* (1823), 317–336; *Journal de Physique, de Chimie, et de l'Histoire Naturelle; par J. C. de Lamétherie et Ducrotay de Blainville, 89* (1819), 210–217.
(R. S. C. 38).

244.
[Anon.], 'Account of Meteoric Stones, Masses of Iron, and Showers of Dust, Red Snow and other substances, which have fallen from the Heavens, from the earliest period down to 1819', *Edinburgh Philosophical Journal, 1* (1819), 221–235. (Editorial by Brewster, possibly with Jameson?)

245.
[Anon.], 'Account of the Experiments of Morichini, Ridolfi, Firmas, and Gibbs, on the Influence of Light in the Development of Magnetism', *Edinburgh Philosophical Journal, 1* (1819), 239–243. (Editorial contribution).

246–256.
[Anon.], 'Historical Account of Discoveries Respecting the Double Refraction and Polarisation of Light', *Edinburgh Philosophical Journal, 1* (1819), 289–296; *ibid., 2* (1820), 167–171; *ibid., 3* (1820), 148–154, 277–285; *ibid., 4* (1820–21), 124–130; *ibid., 8* (1823), 149–160, 245–256; *ibid., 9* (1823), 148–152; *ibid., 10* (1824), 328–331; *Edinburgh Journal of Science, 1* (1824), 77–83, 90–96.
(R. S. C. 73).

257.
[Anon.], 'Account of some Important Discoveries in Magnetism, recently made by P. Barlow, Esq', *Edinburgh Philosophical Journal, 1* (1819), 344–348. (Editorial contribution).

258–259.
'On the Phosphorescence of Minerals', *Edinburgh Philosophical Journal, 1* (1819), 383–388; *Annales de Chimie, ou Recueil de Mémoires concernant la Chimie et les Arts qui en dépendent, 14* (1820), 288–299.
(R. S. C. 39).

260–263.
'On the Laws which Regulate the Absorption of Polarised Light by Doubly Refracting Crystals.' *Philosophical Transactions of the Royal Society of London, 109* (1819), 11–28; *Edinburgh Philosophical Journal, 2* (1820), 341–348; *Annalen der Physik; von L. W. Gilbert, 65* (1820), 4–19; *Journal de Physique, de Chimie, et de l'Histoire Naturelle; par J. C. de Lamétherie et Ducrotay de Blainville, 89* (1819), 210–217.
(R. S. C. 40).

264.
'On the Action of Crystallised Surfaces upon Light', *Philosophical Transactions of the Royal Society of London, 109* (1819), 145–160.
(R. S. C. 41).

265–268.
'On the Optical and Physical Properties of Tabasheer', *Philosophical Transactions of the Royal Society of London, 109* (1819), 283–299; *Edinburgh Philosophical Journal, 1* (1819), 147–150; *ibid., 2* (1820), 97–102; *Journal für Chemie und Physik; von J. S. C. Schweigger, 29* (1820), 411–429.
(R. S. C. 42).

269–272.
A Treatise on the Kaleidoscope (Edinburgh, 1819).
—*The Kaleidoscope: its History, Theory, and Construction, with its Application to the Fine and Useful Arts* . . . 2nd edition [of *A Treatise on the Kaleidoscope*] (London, 1858).
—Another edition (London, 1870).
—Another edition, greatly enlarged (London, 1870).
(B. M. Cat., Nat. Un. Cat.).

273.
[Anon.], 'Captain Ross, and Sir James Lancaster's Sound', *Blackwood's Edinburgh Magazine, 5* (1819), 150–151.
(Strout).

274.
[Anon.], 'Account of Mr. Fresnel's Discoveries Respecting the Inflexion of Light', *Edinburgh Philosophical Journal, 2* (1820), 150–153.
(Morse).

275.
'On a singular development of Crystalline Structure by Phosphorescence', *Edinburgh Philosophical Journal, 2* (1820), 171–172.
(R. S. C. 43).

276–277.
'On the Optical Properties and Mechanical Condition of Amber', *Edinburgh Philosophical Journal, 12* (1820), 332–334; *Annalen der Physik; von L. W. Gilbert, 65* (1820), 20–25.
(R. S. C. 44).

278–279.
'Account of some Single Microscopes upon a New Construction', *Edinburgh Philosophical Journal, 3* (1820), 74–77; *Annales Générales des Sciences Physiques; par Bory de St. Vincent, Drapiez et Van Mons, 6* (1820), 395–398.
(R. S. C. 45).

280.
'Notice Respecting a Single Structure in the Diamond', *Edinburgh Philosophical Journal, 3* (1820), 98–100.
(R. S. C. 46).

281.
'Notice Respecting some New Species of Lead-Ore from Wanlockhead and Lead-hills', *Edinburgh Philosophical Journal, 3* (1820), 138–140.
(R. S. C. 47).

282.
'On the Phenomena of Dichromism on the Absorption of Common Light by Crystallised Bodies', *Edinburgh Philosophical Journal, 3* (1820), 243–249.
(R. S. C. 48).

283–284.
'On a Singular Luminous Property of Wood, etc., steeped in Solutions of Lime and Magnesia', *Edinburgh Philosophical Journal, 3* (1820), 343–344; *Annales Générales des Sciences Physiques; par Bory de St. Vincent, Drapiez et Van Mons, 6* (1820), 260–261.
(R. S. C. 49).

285–286.
[Anon.], 'Remarks on Professor Hansteen's "Inquiries Concerning the Magnetism of the Earth"', *Edinburgh Philosophical Journal, 3* (1820), 124–138; *ibid., 4* (1820), 114–124. (?Editorial contribution).

287–288.
'Additional Observations on the Connection between the Primitive Forms of Minerals and Number of their Axes of Double Refraction', *Journal de Physique, de Chimie, et de l'Histoire Naturelle; par J. C. de Lamétherie et Ducrotay de Blainville, 91* (1820), 300–309; *Memoirs of the Wernerian Natural History Society, 3* (1821), 337–350.
(R. S. C. 50, Morse).

289–290.
'Account of Comptonite, a New Mineral from Vesuvius', *Edinburgh Philosophical Journal, 4* (1821), 131–133; *Journal für Chemie und Physik; von J. S. C. Schweigger, 33* (1821), 278–281.
(R. S. C. 51).

91.
'Description of a New Double Image Micrometer for Measuring the Distance of Celestial Objects', *Edinburgh Philosophical Journal, 4* (1821), 164–167.
(R. S. C. 52).

292.
'Reply to a Note in the *Annales de Chimie* by M. Arago, on the Phosphorescence of Fluor-Spar', *Edinburgh Philosophical Journal, 4* (1821), 180–185.
(R. S. C. 53).

293–294.
'Account of the Native Hydrate of Magnesia Discovered by Dr. Hibbert in Shetland', *Edinburgh Philosophical Journal, 4* (1821), 352–355; *Transactions of The Royal Society of Edinburgh, 9* (1823), 239–242.
(R. S. C. 54).

295.
[Anon.], 'Notice Respecting Professor Hansteen's Chart of the Variation and Dip of the Needle', *Edinburgh Philosophical Journal, 4* (1821), 363–364. (Editorial contribution).

296–298.
'On the connection between the Optical Structure and Chemical Composition of Minerals', *Edinburgh Philosophical Journal, 5* (1821), 1–8; *Annalen der Physik; von L. W. Gilbert, 69* (1821), 157–167; *Journal für Chemie und Physik; von J. S. C. Schweigger, 33* (1821), 340–347.
(R. S. C. 56).

299.
'Account of the Atush-Kudda, or Natural Fire Temples of the Guebres, formed by burning springs of Naphtha, with a notice respecting the Naphtha wells in Pegu', *Edinburgh Philosophical Journal, 5* (1821), 21–27.
(R. S. C. 55).

300.
'Account of the New Galvano-magnetic Condenser, invented by M. Poggendorff of Berlin', *Edinburgh Philosophical Journal, 5* (1821), 112–113. (Signed article).

301.
[Anon.], 'Analysis of Mr. Barlow's Essay on Magnetic Attractions', *Edinburgh Philosophical Journal, 5* (1821), 261–268. (Editorial contribution).

302.
'On the Form of the Integrant Molecule of Carbonate of Lime', *Transactions of the Geological Society of London,* 5 (1821), 83–86.
(R. S. C. 57).

303.
'Observations on Vision through Coloured Glasses, and on their Applications to Telescopes and Microscopes of Great Magnitude', *Edinburgh Philosophical Journal,* 6 (1822), 102–107.
(R. S. C. 58).

304.
'Description of a Teinoscope for Altering the Lineal Proportions of Objects; with Observations on Professor Amici's *Memoir* on Telescopes without Lenses', *Edinburgh Philosophical Journal,* 6 (1822), 334–338.
(R. S. C. 59).

305.
'Observations on the Relation between the Optical Structure and the Chemical Composition of the Apophyllite and other Minerals of the Zeolite Family in Reference to the Preceding Analysis of M. Berzelius', *Edinburgh Philosophical Journal,* 7 (1822), 12–18.
(R. S. C. 60).

306.
'Observations on the Preceding Paper [J. W. F. Herschel on The Law of the Extraordinary Refraction of Differently Coloured Rays]', *Edinburgh Philosophical Journal,* 7 (1882), 142–144. (Signed article).

307.
'Account of a Singular Experiment Depending on the Polarisation of Light by Reflexion', *Edinburgh Philosophical Journal,* 7 (1822), 146–147.
(R. S. C. 61).

308.
D. B., 'Laws of Polarisation in Rectangular Plates of Glass', *Edinburgh Philosophical Journal,* 7 (1822), 178.
(Morse).

309.
'Description of a New Reflecting Telescope', *Edinburgh Philosophical Journal,* 7 (1822), 323–328.
(R. S. C. 62).

310. John Robison
A System of Mechanical Philosophy; with notes by D. Brewster (Edinburgh, 1822).
(B. M. Cat.).

311–312.
'On the Construction of Polyzonal Lenses and Mirrors of Great Magnitude for Lighthouses and for Burning Instruments, and on the Formation of a Great National Burning Apparatus', *Edinburgh Philosophical Journal,* 8 (1823), 160–169; *Transactions of the Royal Society of Edinburgh,* 9 (1831), 33–72.
(R. S. C. 63).

313.
'Description of a New Reflecting Microscope', *Edinburgh Philosophical Journal,* 8 (1823), 326–327.
(R. S. C. 64).

314–321.
'On the Existence of Two New Fluids in the Cavities of Minerals, which are Immiscible, and Possess Remarkable Physical Properties', *Edinburgh Philosophical Journal,* 9 (1823), 94–95, 268–270, 400; *Transactions of the Royal Society of Edinburgh,* 10 (1826), 1–42; *Annales de Chimie, ou Recueil de Mémoires concernant la Chimie et les Arts qui en dépendent,* new series, 23 (1823), 305–310; *ibid.,* 25 (1824), 75–78; *Annalen der Physik und Chemie; von J. C. Poggendorff,* 7 (1826), 469–488; *The American Journal of Science and Art; by B. Silliman,* 12 (1827), 214–227; *Journal für Chemie und Physik; von J. S. C. Schweigger,* 38 (1823), 229–230; *ibid.,* 40 (1824), 177–199.
(R. S. C. 65).

322.
[Anon.], 'On the Knight's Moves over the Chessboard', *Edinburgh Philosophical Journal,* 9 (1823), 236–237.
(R. S. C. 66).

323.
'On the Existence of a Group of Moveable Crystals of Carbonate of Lime in a Fluid Cavity of Quartz', *Edinburgh Philosophical Journal,* 9 (1823), 268–270. (Signed article).

324.
'Reply to Mr. Brooke's Observations on the Connection between the Optical Structure of Minerals and their Primitive Forms', *Edinburgh Philosophical Journal,* 9 (1823), 361–372.
(R. S. C. 67).

325.
'On Circular Polarisation as Exhibited in the Optical Structure of the Amethyst, with Remarks on the Distribution of the Colouring Matter of that Mineral', *Transactions of the Royal Society of Edinburgh,* 9 (1823), 139–152.
(R. S. C. 68).

326–328.
'Observations on the Mean Temperature of the Globe', *Transactions of the Royal Society of Edinburgh,* 9 (1823), 201–226; *Edinburgh Journal of Science,* new series, 4 (1831), 300–320; *Notizen aus dem Gebiete der Natur und Heilkunde; von L. Froriep,* 30 (1831), 225–234.
(R. S. C. 69).

329–331.
'Description of a Monochromatic Lamp for Microscopical Purposes, with Remarks on the Absorption of the Prismatic Rays by Coloured Media', *Transactions of the Royal Society of Edinburgh,* 9 (1823), 433–444; *Edinburgh Philosophical Journal,* 10 (1824), 120–125; *Annalen der Physik und Chemie; von J. C. Poggendorff,* 2 (1824), 98–106.
(R. S. C. 70).

332–338. Leonhard Euler
Letters to a German Princess on different subjects in Physics and Philosophy. Translated from the French by Henry Hunter, D.D., with notes and a life of Euler by David Brewster, LL.D. (Edinburgh and London, 1823).
—American edition, with additional notes by J. Griscom 2 vols. (New York, 1833).
—Another edition (New York, 1837).
—Another edition (New York, 1838).
—Another edition (New York, 1840).
—Another edition (New York, 1842).
—Facsimile edition, from the 1833 edition (New York, 1975).
(Gordon, B. M. Cat., Nat. Un. Cat., Books in Print).

339. James Ferguson
Essays and Treatises on Astronomy, Drawing in Perspective, Electricity &c.; with an Appendix relative to Electricity, Galvanism, and Electro-Magnetism, by David Brewster. (Edinburgh, 1823).
(Signet Library).

340.
[Anon.], 'Hydrodynamics' *Encyclopaedia Britannica,* 20 vols., 6th edition (Edinburgh, 1820–1823), X, 697–792.
(List).

341.
[Anon.], 'Mathematics (history)' *Encyclopaedia Britannica,* 20 vols., 6th edition (Edinburgh, 1820–1823), XIII, 1–14.
(List).

342.
[Anon.], 'Mechanics', *Encyclopaedia Britannica,* 20 vols., 6th edition (Edinburgh, 1820–1823), XIII, 47–136.
(List).

343.
[Anon.], 'Optics' *Encyclopaedia Britannica*, 20 vols., 6th edition (Edinburgh, 1820–1823), XV, 171–295.
(List).

344.
[Anon.], 'Telescope' *Encyclopaedia Britannica*, 20 vols., 6th edition (Edinburgh, 1820–1823), XX, 236–272.
(List).

345–347.
'On a new species of Double Refraction, accompanying a remarkable structure in Analcime', *Edinburgh Philosophical Journal*, 10 (1824), 255–259; *Transactions of the Royal Society of Edinburgh*, 10 (1826), 187–194; *Zeitschrift für Physik, Mathematic, und verwandte Wissenschaften; von A. Baumgartner und A. von Ettingschausen*, 2 (1827), 21–29.
(R. S. C. 71).

348–349.
'On the Accommodation of the Eye to Different Distances', *Edinburgh Journal of Science*, 1 (1824), 77–83; *Annalen der Physik und Chemie; von J. C. Poggendorff*, 2 (1824), 271–281.
(R. S. C. 72).

350–351.
[Anon.], 'Tables of the Variations of the Magnetic Needle in Different Parts of the Globe', *Edinburgh Journal of Science*, 1 (1824), 87–90, 334–335. (? Editorial contribution).

352–353.
'Description of Two Surfaces composed of Siliceous Filaments incapable of Reflecting Light, and produced by the Fracture of a Large Piece of Quartz', *Edinburgh Journal of Science*, 1 (1824), 108–110; *Annalen der Physik und Chemie; von J. C. Poggendorff*, 2 (1824), 293–297.
(R. S. C. 74).

354–357.
'Observations on the Pyro-Electricity of Minerals', *Edinburgh Journal of Science*, 1 (1824), 208–215; *Annales de Chimie, ou Recueil de Mémoires concernant la Chimie et les Arts qui en dépendent*, 28 (1825), 161–164; *Annalen der Physik und Chemie; von J. C. Poggendorff*, 2 (1824), 297–308; *Journal für Chemie und Physik; von J. S. C. Schweigger*, 43 (1825), 87–106.
(R. S. C. 75).

358–362.   Adrien Marie Legendre
Elements of Geometry and Trigonometry; with notes. Translated from the French by T. Carlyle. Edited by D. Brewster. With notes and additions by the author, and an introductory chapter on Proportion by the translator (Edinburgh, 1824).
—Another edition, revised and adapted to the course of mathematical instruction in the United States, by C. Davies (New York, 1834).
—Another edition (New York, 1835).
—Another edition (New York and Boston, 1835).
—Another edition (New York, 1846).
(B. M. Cat., Nat. Un. Cat.).

363.
'Observations on the Vision of Impressions on the Retina, in Reference to Certain Supposed Discoveries respecting Vision announced by Mr. Charles Bell', *Edinburgh Journal of Science*, 2 (1825), 1–9.
(R. S. C. 76).

364.
'On the formation of single Microscopes from the Lenses of Fishes etc', *Edinburgh Journal of Science*, 2 (1825), 98–100.
(R. S. C. 77).

365.
[Anon.], 'Description of an extraordinary Parhelion observed at Gotha, 12th May 1824', *Edinburgh Journal of Science*, 2 (1825), 105–107.
(R. S. C. 78).

366–367.
'On the structure of Rice Paper', *Edinburgh Journal of Science*, 2 (1825), 135–136; *Journal für Chemie und Physik; von J. S. C. Schweigger*, 45 (1825), 247–248.
(R. S. C. 79).

368.
[Anon.], 'On the Convergency of the Solar Beams to a point opposite the Sun', *Edinburgh Journal of Science*, 2 (1825), 136–138.
(R. S. C. 300).

369.
[Anon.], 'On the Existence of Siliceous Solutions in the Drusy Cavities of Minerals', *Edinburgh Journal of Science*, 2 (1825), 140–142. (Corrects mistaken reading of an earlier Brewster paper).

370.
'Observations on the Optical Structure of Lithion-Mica, analysed by Professor Gmelin', *Edinburgh Journal of Science*, 2 (1825), 205–206.
(R. S. C. 80).

371.
'Description of Withamite, a New Mineral Species found in Glenco', *Edinburgh Journal of Science*, 2 (1825), 218–221.
(R. S. C. 81).

372.
'Description of Gmelinite, a New Mineral Species', *Edinburgh Journal of Science*, 2 (1825), 262–267.
(R. S. C. 82).

373.
[Anon.], 'Description of Fraunhöfer's large Achromatic Telescopes: with a Plate', *Edinburgh Journal of Science*, 2 (1825), 305–307. (Editorial contribution).

374.
'Description of Levyne, a New Mineral Species', *Edinburgh Journal of Science*, 2 (1825), 332–334; *ibid.*, 4 (1826), 316–317.
(R. S. C. 83).

375.
[Anon.], 'Account of the Explosions of Oil Gas which took place at Edinburgh on the 23rd March 1825, with Observations on the Safety of Gas', *Edinburgh Journal of Science*, 3 (1825), 83–93.
(? Editorial contribution).

376–377.
'On some Remarkable Affections of the Retina, as Exhibited in its Insensibility to Indirect Impressions, and to the Impressions of Attenuated Light', *Edinburgh Journal of Science*, 3 (1825), 288–293; *Report of the British Association for the Advancement of Science . . . 1832* (London, 1833), 551–553.
(R. S. C. 84).

378.
[Anon.], 'Biographical notice of the Abbé Haüy', *Edinburgh Journal of Science*, 4 (1826), 1–12. (Editorial additions to Cuvier's 'Eloge').

379.
[Anon.], 'A Popular Summary of the Experiments of Messrs. Barlow, Christie, Babbage, and Herschel on the Magnetism of Iron and Other Metals, as Exhibited by Rotation', *Edinburgh Journal of Science*, 4 (1826), 13–19. (Editorial contribution).

380.
D. B., 'On the Optical Illusion of the Conversion of Cameos into Intaglios, and of Intaglios into Cameos, with an Account of other Analogous Phenomena', *Edinburgh Journal of Science*, 4 (1826), 99–108.
(R. S. C. 85).

381.
[Anon.], 'Notice of Captain Parry's Last Expedition to the Arctic Regions in 1824 and 1825', *Edinburgh Journal of Science*, 4 (1826), 147–150. (Editorial contribution).

382.
[Anon.], 'Further Observations on the History of Mr. Babbage's Experiments on the Production of Magnetism by Rotation', *Edinburgh Journal of Science*, 4 (1826), 257–260. (Editorial contribution).

383.
D. B., 'Observations on the Superiority of Achromatic Telescopes with Fluid Object-Glasses, as constructed by Dr. Blair', *Edinburgh Journal of Science*, 4 (1826), 282–285.

384.
[Anon.], 'Observations on the Relative Performances of Achromatic and Reflecting Telescopes by Mr. Herschel, Mr. Smith, and Professor Amici', *Edinburgh Journal of Science*, 4 (1826), 309–315. (Editorial contribution).

385.
[Anon.], 'Biographical Memoir of Mark Augustus Pictet', *Edinburgh Journal of Science*, 5 (1826), 1–8. (Editorial contribution).

386–387.
'Results of the Thermometrical Observations made at Leith Fort Every Hour of the Day and Night during the Whole of the Years 1824–1825', *Transactions of the Royal Society of Edinburgh*, 10 (1826), 326–388; *Edinburgh Journal of Science*, 5 (1826), 18–32.
(R. S. C. 86).

388.
[Anon.], 'Observations Relative to the Sound which accompanies the Aurora Borealis', *Edinburgh Journal of Science*, 5 (1826), 74–77. (Editorial contribution).

388–395.
'On the Refractive Power of the Two New Fluids in Minerals, with Additional Observations on the Nature and Properties of these Substances', *Transactions of the Royal Society of Edinburgh*, 10 (1826), 407–427; *Zeitschrift für Physik, Mathematik, und verwandte Wissenschaften; von A. Baumgartner und A. von Ettingshausen*, 1 (1826), 414–430; *Edinburgh Journal of Science*, 5 (1826), 122–136; *Notizen aus dem Gebiete der Natur und Heilkunde; von L. von Froriep*, 15 (1826), 49–55, 64–72; *Annalen der Physik und Chemie; von J. C. Poggendorff*, 7 (1826), 489–514; *Journal für Chemie und Physik; von J. S. C. Schweigger*, 47 (1826), 213–236.
(R. S. C. 87).

396.
[Anon.], 'Further Observations on the Supposed Optical and Physiological Discoveries of Mr. Charles Bell', *Edinburgh Journal of Science*, 5 (1826), 259–268. (Editorial contribution).

397.
[Anon.], 'Description of an Apparatus for Producing Intense Light, Visible at Great Distances, invented by Lieutenant Thomas Drummond R.E.', *Edinburgh Journal of Science*, 5 (1826), 319–322. (? Editorial contribution).

398.
'Description of Hopeite, a New Mineral from Altenberg, near Aix-la-Chapelle', *Transactions of the Royal Society of Edinburgh*, 10 (1826), 107–111.
(R. S. C. 88).

399.
'On the Distribution of the Colouring Matter, and on Certain Peculiarities in the Structure and Properties of the Brazilian Topaz', *Transactions of the Cambridge Philosophical Society*, 2 (1827), 1–10.
(R. S. C. 89).

400.
'Notice Respecting the Mean Temperature of the Equator', *Edinburgh Journal of Science*, 6 (1827), 117–120.
(R. S. C. 90).

401.
[Anon.], 'Notice Respecting the Hourly Meteorological Observations proposed by the Royal Society of Edinburgh to be made Twice Every Year on the 17th July and 15th January', *Edinburgh Journal of Science*, 6 (1827), 144–149. (Editorial contribution).

402.
[Anon.], 'On the Subterranean Sounds heard at Nakous, on the Red Sea', *Edinburgh Journal of Science*, 6 (1827), 153–155. (? Editorial contribution).

403–404.
'Notice respecting the Existence of the New Fluid in a Large Cavity in a Specimen of Sapphire', *Edinburgh Journal of Science*, 6 (1827), 155–156; *Annalen der Physik und Chemie; von J. C. Poggendorff*, 9 (1827), 510–512.
(R. S. C. 91).

405.
[Anon.], 'Memoir of the Life and Writings of M. Piazzi, Director General of the Observatories of Naples and Palermo', *Edinburgh Journal of Science*, 6 (1827), 193–199. (Editorial contribution).

406.
'On the Separation of Epistilbite from Heulandite, as Demonstrated by Optical Characters', *Edinburgh Journal of Science*, 6 (1827), 236–240.
(R. S. C. 92).

407.
'Account of an Improvement in the Nautical Eye-Tube', *Edinburgh Journal of Science*, 6 (1827), 250–251.
(R. S. C. 93).

408.
[Anon.], 'On the Different Primitive Forms of the same Salt produced by a change in the Nature of the Solvent by Dr. Christian Wollner: with observations by the Editor', *Edinburgh Journal of Science*, 6 (1827), 289–292. (Editorial contribution).

409.
'Observations on the Structure and Crystalline Forms of Haytorite: in a Letter from Dr. Brewster to Cornelius Tripe Esq., Devonport', *Edinburgh Journal of Science*, 6 (1827), 301–307.
(R. S. C. 94).

410–411.
[Anon.], 'On the Systems of Double Stars, which have been Demonstrated to be Binary ones, by the Observations of Sir W. Herschel, and Messrs. Herschel and South', *Edinburgh Journal of Science*, 7 (1827), 88–97; *ibid.*, 8 (1828), 40–44.
(R. S. C. 95).

412.
'Description of Oxahverite, a New Mineral from Oxahver, in Iceland', *Edinburgh Journal of Science*, 7 (1827), 115–118.
(R. S. C. 96).

413.
[Anon.], 'Notice respecting Professor Barlow's new Achromatic Telescopes with Fluid Object-Glasses', *Edinburgh Journal of Science*, 7 (1827), 335–336.
(R. S. C. 97).

414.
'A Table of Refractive Densities', *Quarterly Journal of Science*, 22 (1827), 355–365.
(R. S. C. 98).

415.
[Anon.], Review of J. Fraunhöfer's 'Refractive and Dispersive Powers of Glass and the Achromatic Telescope', *Foreign Quarterly Review*, 1 (1827), 424–435.
(W. II.).

416.
[Anon.], 'Biographical Sketch of Alexander Volta, Professor of

Natural Philosophy at Como', *Edinburgh Journal of Science, 8* (1828), 1–6. (Editorial contribution).

417.
'Account of the Fossil Bones found on the left bank of the Irrawadi, in Ava', *Edinburgh Journal of Science, 8* (1828), 56–60.
(R. S. C. 99).

418–419.
'On the Mean Temperature of the Equator, as deduced from Observations made at Prince of Wales's Island, Singapore, and Malacca', *Edinburgh Journal of Science, 8* (1828), 60–67; *Notizen aus dem Gebiete der Natur und Heilkunde; von L. von Froriep, 24* (1829), 145–149.
(R. S. C. 100).

420.
[Anon.], 'Notice Respecting the Varnish and Varnish Trees of India', *Edinburgh Journal of Science, 8* (1828), 96–100.
(R. S. C. 101).

421.
[Anon.], 'Account of the Poisonous Qualities of the Vegetable Varnishes of America and India', *Edinburgh Journal of Science, 8* (1828), 100–104.
(R. S. C. 102).

422–423.
[Anon.], 'On the Supposed Influence of the Aurora Borealis upon the Magnetic Needle, in Reply to the Observations of M. Arago, as Communicated to the Academy of Science on the 22nd January 1828', *Edinburgh Journal of Science, 8* (1828), 189–201; *Notizen aus dem Gebiete der Natur und Heilkunde; von L. von Froriep, 21* (1828), 145–153.
(R. S. C. 103).

424–426.
'On the Natural History and Properties of Tabasheer, the Siliceous Concretion in the Bamboo', *Edinburgh Journal of Science, 8* (1828), 285–294; *Notizen aus dem Gebiete der Natur und Heilkunde; von L. von Froriep, 21* (1828), 179–186; *Journal für Chemie und Physik; von J. S. C. Schweigger, 52* (1828), 412–426.
(R. S. C. 104).

427.
[Anon.], 'Account of the Performances of Different Ventriloquists, with Observations on the Art of Ventriloquism', *Edinburgh Journal of Science, 9* (1828), 252–259.
(R. S. C. 105).

428.
'On a New Cleavage in Calcareous Spar, with a Notice of a Method of Detecting Secondary Cleavages in Minerals', *Edinburgh Journal of Science, 9* (1828), 311–314.
(R. S. C. 106).

429.
[Anon.], 'Recent History of Astronomy', *Quarterly Review, 38* (1828), 1–15.
(W. I.).

430.
[Anon.], 'Facts and Observations Relative to the Recent Formation of Quartz Crystals etc., and of Indurated Calcedony from Siliceous Solutions and Paste', *Edinburgh Journal of Science, 10* (1829), 28–33.
(R. S. C. 107).

431.
[Anon.], 'Account of Two Remarkable Cases of Insensibility in the Eye to Particular Colours', *Edinburgh Journal of Science, 10* (1829), 153–159.
(R. S. C. 108).

432.
'Account of Two Remarkable Rainbows, one of which Enclosed the Phenomenon of Converging Solar Beams', *Edinburgh Journal of Science, 10* (1829), 163–164.
(R. S. C. 109).

433.
[Anon.], 'Biographical Sketch of the late Dugald Stewart Esq., F.R.S. Lond. and Ed.', *Edinburgh Journal of Science, 10* (1829), 193–206. (Editorial contribution).

434.
'Observations relative to the Motions of the Molecules of Bodies', *Edinburgh Journal of Science, 10* (1829), 215–220.
(R. S. C. 110).

435–438.
'Account of a Remarkable Peculiarity in the Structure of Glauberite, which has one Axis of Double Refraction for Violet, and Two Axes for Red Light', *Edinburgh Journal of Science, 10* (1829), 325–327; *Journal für Chemie und Physik; von J. S. C. Schweigger, 55* (1829), 318–321; *Transactions of the Royal Society of Edinburgh, 11* (1831), 273–276; *Annalen der Physik und Chemie; von J. C. Poggendorff, 21* (1831), 607–609.
(R. S. C. 111).

439–440.
[Anon.], 'Account of the Single Lens Microscope of Sapphire and Diamond constructed by Mr. A. Pritchard', *Edinburgh Journal of Science, 10* (1829), 327–333; *Annalen der Physik und Chemie; von J. C. Poggendorff, 15* (1829), 517–522.
(R. S. C. 112).

441–442.
'Notice Respecting a Method of Producing an Intense Heat from Gas for Various Purposes in the Arts', *Edinburgh Journal of Science*, new series, *1* (1829), 104–108; *Journal für Chemie und Physik; von J. S. C. Schweigger, 55* (1829), 364–369.
(R. S. C. 113).

443–445.
'Account of a new Monochromatic Lamp depending on the Combustion of Compressed Gas', *Edinburgh Journal of Science*, new series, *1* (1829), 108; *Annalen der Physik und Chemie; von J. C. Poggendorff, 16* (1829), 379–383; *Journal für Chemie und Physik; von J. S. C. Schweigger, 55* (1829), 369–376.
(R. S. C. 114).

446–449.
'On the Reflexion and Decomposition of Light at the Separating Surfaces of Media of the same and of different Refractive Powers', *Philosophical Transactions of the Royal Society of London, 119* (1829), 187–206; *Edinburgh Journal of Science*, new series, *1* (1829), 209–229; *Annalen der Physik und Chemie; von J. C. Poggendorff, 17* (1829), 29–53; *Journal für Chemie und Physik; von J. S. C. Schweigger, 55* (1829), 115–182.
(R. S. C. 115).

450–453.
'On a new Series of Periodical Colours produced by the Grooved Surfaces of Metallic and Transparent Bodies', *Philosophical Transactions of the Royal Society of London, 119* (1829), 301–316; *Edinburgh Journal of Science*, new series, *2* (1830), 46–61; *Notizen aus dem Gebiete der Natur und Heilkunde; von L. von Froriep, 27* (1830), 145–154; *Annalen der Physik und Chemie; von J. C. Poggendorff, 18* (1830), 579–596.
(R. S. C. 116).

454.
[Anon.], 'Dr. Lloyd on Mechanical Philosophy', *Quarterly Review, 39* (1829), 432–451.
(W. I.).

455.
[Anon.], 'Systems and Methods in Natural History', *Quarterly Review, 41* (1829), 302–327.
(W. I.).

456.
[Anon.], 'Obituary: J. von Fraunhöfer', *American Journal of Science and Arts; by B. Silliman*, 16 (1829), 301.
(Sutton).

457–458.
[Anon.], Articles 'Optics', and 'Double Refraction and Polarisation of Light', in *Natural Philosophy*, 4 vols., (London, 1829–1834), I, Parts 7 and 8; in *Library of Useful Knowledge* published by the Society for the Diffusion of Useful Knowledge.
(B. M. Cat., Morse).

459.
'On the Law of Partial Polarisation of Light by Reflection', *Edinburgh Journal of Science*, new series, 3 (1830), 160–177.
(R. S. C. 117).

460–466.
'On the Laws of Polarisation of Light by Refraction', *Edinburgh Journal of Science*, new series, 3 (1830), 218–230; *Philosophical Transactions of the Royal Society of London*, 120 (1830), 69–84, 133–144; *Annalen der Physik und Chemie; von J. C. Poggendorff*, 19 (1830), 259–280, 281–294; *The American Journal of Science and Arts; by B. Silliman*, 22 (1832), 277–292; ibid., 23 (1833), 225–236.
(R. S. C. 120).

467–470.
'On the action of the Second Surfaces of Transparent Plates upon Light', *Edinburgh Journal of Science*, new series, 3 (1830), 230–236; *Philosophical Transactions of the Royal Society of London*, 120 (1830), 145–152; *Annalen der Physik und Chemie; von J. C. Poggendorff*, 19 (1830), 518–526; *The American Journal of Science and Arts; by B. Silliman*, 23 (1833), 28–35.
(R. S. C. 121).

471.
[Anon.], 'Account of a Fifth Case of Spectral Illusion', *Edinburgh Journal of Science*, new series, 3 (1830), 244–245. (? Editorial contribution).

472.
'Microscopical Examination of the Peculiar Structure observed in the Second Stomach of certain Cetacea', *Edinburgh Journal of Science*, new series, 3 (1830), 327–328.
(R. S. C. 118).

473–476.
'On the Production of Regular Double Refraction in the Molecules of Bodies by Simple Pressure; with Observations on the Origins of the Doubly Refracting Structure', *Philosophical Transactions of the Royal Society of London*, 120 (1830), 87–96; *Edinburgh Journal of Science*, new series, 3 (1830), 328–337; *Annalen der Physik und Chemie; J. C. Poggendorff*, 19 (1830), 527–539; *The American Journal of Science and Arts; by B. Silliman*, 21 (1832), 296–304.
(R. S. C. 119).

477–480.
'On the Phenomena and Laws of Elliptic Polarisation, as Exhibited in the Action of Metal Plates upon Light', *Philosophical Transactions of the Royal Society of London*, 120 (1830), 287–326; *Edinburgh Journal of Science*, new series, 4 (1831), 136–165, 247–261; *Annalen der Physik und Chemie; von J. C. Poggendorff*, 21 (1831), 219–275.
(R. S. C. 122).

481–482.
Edinburgh Encyclopaedia, 20 vols., (Edinburgh, 1830).
—21 vols, 1st American edition (Philadelphia, 1832).
Major contributions by the Editor.

483.
(w), 'Accidental Colours', I, 88–93.
(List).

484.
(w), 'Achromatic Telescope', I, 104–110.
(List).

485.
(β), 'Alembert, John Le Rond D", I, 398–402.
(List).

486.
(β), 'Almamon [Al-Mamon]', I, 543–545.
(List).

487.
(β), 'Anemometer', II, 69–77.
(List).

488.
(β), 'Astronomy, History', II, 582–606.
(List).

489.
(β), 'Astronomy, Descriptive', II, 606–680.
(List).

490.
(β), 'Astronomy, Practical', II, 720–832.
(List).

491.
(o), 'Atmosphere, Physical', III, 55–57.
(Whewell).

492.
(β), 'Bailly, Jean Sylvain', III, 204–207.
(List).

493.
(β), 'Bernouilli, James', III, 479–482.
(List).

494.
(β), 'Bernouilli, John', III, 482–484.
(List).

495.
(β), 'Bernouilli, Daniel', III, 484–486.
(List).

496.
(β), 'Boscovitch, Roger Joseph', III, 744–749.
(List).

497.
(β), 'Bradley, James', IV, 397–399.
(List).

498.
(β), 'Brahe, Tycho', IV, 400–403.
(List).

499.
(β), 'Buffon, George Louis Le Clerc', V, 76–81.
(List).

500.
(β), 'Burning Instruments', V, 130–144.
(List).

501.
(β), 'Capilliary Attraction', V, 406–412.
(Whewell).

502.
(o), 'Condamine, Charles Maria de la', VII, 110–112.
(List).

503.
(o), 'Condorcet, Jean Antoine Nicolas de Caritat, Marquis de', VII, 113–115.
(List).

504.
(β), 'Copernicus, (Nicolas), or Zepernick', VII, 207–209.
(List).

505.
(β), 'Electricity', VIII, 411–550.
(List).

506.
(β), 'Euler, Leonard', IX, 228–231.
(List).

507.
(β), 'Expansion', IX, 254–261.
(List).

508.
(β), 'Galileo Galilei', X, 73–76.
(List).

509.
[Anon.], 'Goniometer', X, 336–338.
(Whewell).

510.
[Anon.], 'Gregory, James', X, 506–508.
(List).

511.
[Anon.], 'Gregory, David, Dr.', X, 508–510.
(List).

512.
[Anon.], 'Gregory, John, Dr.', X, 510–511.
(List).

513.
[Anon.], 'Halley, Edmund', X, 609–612.
(List).

514.
[Anon.], 'Hydrodynamics', XI, 408–568.
(List).

515.
[Anon.], 'Kaleidoscope', XII, 410–412.
(List).

516.
[Anon.], 'Mechanics', XIII, 500–623.
(List).

517.
[Anon.], 'Micrometer', XIV, 198–215.
(List).

518.
[Anon.], 'Microscope', XIV, 215–232.
(List).

519.
[Anon.], 'Optics', XV, 460–662.
(List).

520.
[Anon.], 'Steam', XVIII, 348–354.
(Whewell).

521.
[Anon.], 'Steam Engine', XVIII, 354–381.
(Whewell).

522.
[Anon.], 'Steam-Boat', XVIII, 381–383.
(Whewell).

523.
[Anon.], 'Steam Carriage', XVIII, 383–384.
(Whewell).

524.
[Anon.], 'Decline of Science in England and Patent Laws', *Quarterly Review*, 43 (1830), 305–342.
(W. I.).

525.
[Anon.], 'Account of Four other Cases of Spectral Illusion', *Edinburgh Journal of Science*, new series, 4 (1831), 261–263.
(? Editorial contribution).

526.
[Anon.], 'Observations on the Decline of Science in England', *Edinburgh Journal of Science*, new series, 5 (1831), 1–16. (Editorial contribution).

527.
[Anon.], 'Remarks on Dr Goring's Observations on the use of Monochromatic Light with the Microscope', *Edinburgh Journal of Science*, new series, 5 (1831), 143–148.
(R. S. C. 123).

528–530.
'On a New Analysis of Solar Light Indicating the Three Primary Colours, forming Coincident Spectra of Equal Length', *Edinburgh Journal of Science,* new series, 5 (1831), 197–206; *Transactions of the Royal Society of Edinburgh*, 12 (1834), 123–136; *Annalen der Physik und Chemie; von J. C. Poggendorff*, 23 (1831), 435–444.
(R. S. C. 124).

531.
'On the Mean Temperature of Thirty-Four Different Places in the State of New York for 1830', *Edinburgh Journal of Science*, new series, 5 (1831), 255–265.
(R. S. C. 125).

532.
[Anon.], 'Observations on a Pamphlet entitled *On the Alleged Decline of Science in England. By a Foreigner*', *Edinburgh Journal of Science*, new series, 5 (1831), 334–358. (Editorial contribution).

533.
'On the Construction of Polyzonal Lenses, and their Combination with Plain Mirrors, for the Purposes of Illumination in Lighthouses', *Transactions of the Royal Society of Edinburgh*, 11 (1831), 33–72.
(R. S. C. 126).

534.
'On certain new Phenomena of Colour in Labrador Felspar, with Observations on the Nature and Cause of its Changeable Tints', *Transactions of the Royal Society of Edinburgh*, 11 (1831), 322–331.
(R. S. C. 127).

535.
'Ueber die mathematischen Ausdrücke für die mittlere Wärme der Luft und die magnetische Intensität', *Annalen der Physik und Chemie; von J. C. Poggendorff*, 21 (1831), 323–325.
(R. S. C. 128).

536.
[Anon.], 'Connexion of Intellectual Operations with Organic Action', *Quarterly Review*, 45 (1831), 341–358.
(W. I.).

537.
[Anon.], Review of Herschel's 'Treatise on Sound', *Quarterly Review*, 45 (1831), 475–511.
(W. I.).

538–554.
The Life of Sir Isaac Newton (London, 1831).
—Another edition (New York, 1831).

—Another edition (New York, 1833).
—Another edition (New York, 1835).
—Another edition (New York, 1838).
—Another edition (New York, 1839).
—Another edition (New York, 1842).
—Another edition (New York, 1845).
—Another edition (New York, 1848).
—Another edition (New York, 1852).
—Revised and edited by W. T. Lynn (London, 186?).
—Another edition (New York, 1860).
—Another edition (New York, 1864).
—Another edition (New York, 1871).
—Another edition (New York, 1874).
—Another edition (London, 1875).
—*Sir Isaak Newton's Leben*: nebst einer Darstellung seiner Entdeckungen. Uber setzt von B. M. Goldberg, mit Anmerkungen von H. W. Brandes, (Leipzig, 1833).
(B. M. Cat., Nat. Un. Cat.).

555–574.
A Treatise on Optics in *Dr. Lardner's Cabinet Cyclopaedia* (London, 1831).
—1st American edition, ed. A. D. Brache (Philadelphia, 1835).
—2nd American edition, (Philadelphia, 1835).
—Another edition (Philadelphia, 1837).
—Another edition (London, 1838).
—Another edition (Philadelphia, 1838).
—Another edition (Philadelphia, 1839).
—Another edition (Philadelphia, 1841).
—New edition (London, 1843).
—A new edition, with an appendix containing an elementary view of the application and analysis to reflexion and refraction, by A. D. Brache (Philadelphia, 1844).
—Another edition (Philadelphia, 1845).
—Another edition (Philadelphia, 1847).
—Another edition (Philadelphia, 1848).
—Another edition (London, 1849).
—Another edition (Philadelphia, 1850).
—Another edition (Philadelphia, 1852).
—Another edition (London, 1853).
—Another edition (Philadelphia, 1854).
—*Manual d'optique, ou traité complet et simplifié de cette science* ... Traduit par M. P. Vergnandin Libraire encyclopedique Roret, 2 vols. (Paris, 1833).
—*Populares, voll standiges Handbuch der Optik*. Von D. Brewster ... In's Deutsche ubersetzt von Dr. J. Hartmann (Quedlinburg, 1835).
(B. M. Cat., Morse, Nat. Un. Cat.).

575.
[Anon.], 'On the Direct Encouragement Afforded to Science by the French Government', *Edinburgh Journal of Science*, new series, 6 (1832), 39–43. (Editorial contribution).

576.
'On the Principle of Illumination for Microscopic Objects', *Edinburgh Journal of Science*, new series, 6 (1832), 83–85.
(R. S. C. 132).

577–578.
'Observations on the Preceding Memoir on the Refraction of the Differently Coloured Rays in Crystals with One and Two Axes of Double Refraction by M. Rudberg', *London and Edinburgh Philosophical Magazine and Journal of Science*, 3rd series, 1 (1832), 6–9; 'Observations on the Preceding Memoir, Section II', *ibid.*, 146–147.
(R. S. C. 133).

579.
[Anon.], 'On the Encouragement given to Science by the King of Denmark', *London and Edinburgh Philosophical Magazine and Journal of Science*, 3rd series, 1 (1832), 16–17. (Editorial contribution).

580–582.
'On a New Species of Coloured Fringes, produced from Reflexion between the Lenses of Achromatic Compound Object Glasses', *London and Edinburgh Philosophical Magazine and Journal of Science*, 3rd series, 1 (1832), 19–23; *Transactions of the Royal Society of Edinburgh*, 12 (1834), 191–196; *Annalen der Physik und Chemie; von J. C. Poggendorff*, 26 (1832), 150–156.
(R. S. C. 134).

583–585.
'On the effect of Compression and Dilatation upon the Retina', *London and Edinburgh Philosophical Magazine and Journal of Science*, 3rd series, 1 (1832), 89–92; *Notizen aus dem Gebiete der Natur und Heilkunde; von L. von Froriep*, 36 (1833), 145–148; *Annalen der Physik und Chemie; von J. C. Poggendorff*, 26 (1832), 156–160.
(R. S. C. 135).

586.
D. B., 'Note on the Mean Temperature of Nicolaieff, as deduced from the observations of M. Coumani by Professor M. A. Kupffer: Observations on the preceding results', *London and Edinburgh Philosophical Magazine and Journal of Science*, 3rd series, 1 (1832), 135–136.
(R. S. C. 136).

587.
D. B., 'Note on the Mean Temperature of Sevastopol, as Deduced from the Observations of M. Coumani; by Professor M. A. Kupffer of the Imperial Academy of Sciences of St Petersburg: Observations on the Preceding Results', *London and Edinburgh Philosophical Magazine and Journal of Science*, 3rd series, 1 (1832), 260–261.

588.
D. B., 'On the Variations which Temperature produces in the Double Refraction of Crystals; by Fredrick Rudberg, Professor of Physics in the University of Upsal: Observations on the Preceding Paper', *London and Edinburgh Philosophical Magazine and Journal of Science*, 3rd series, 1 (1832), 415–417.

589–590.
'On the Action of Heat in changing the Number and Nature of the Optical or Resultant Axes of Glauberite', *London and Edinburgh Philosophical Magazine and Journal of Science*, 3rd series, 1 (1832), 417–420; *Annalen der Physik und Chemie; von J. C. Poggendorff*, 27 (1833), 480–484.
(R. S. C. 137).

591–592.
'Observations on the Isothermal Lines on the North West Coast of America, as deduced from the results in the two preceding articles', *London and Edinburgh Philosophical Magazine and Journal of Science*, 3rd series, 1 (1832), 431–432; *Notizen aus dem Gebiete der Natur und Heilkunde; von L. von Froriep*, 35 (1833), 277–278.
(R. S. C. 139).

593–594.
'Account of a Curious Chinese Mirror, which Reflects from its Polished Face the Figures Embossed upon its Back', *London and Edinburgh Philosophical Magazine and Journal of Science*, 3rd series, 1 (1832), 438–441; *Annalen der Physik und Chemie; von J. C. Poggendorff*, 27 (1833), 485–489.
(R. S. C. 138).

595.
[Anon.], 'Portraits of Eminent Philosophers (No I): Sir Isaac Newton', *Fraser's Magazine*, 6 (1832), 351–359.
(W. II.).

596.
[Anon.], 'Portraits of Eminent Philosophers (No II, concl.): Marquess de la Place', *Fraser's Magazine*, 6 (1832), 446–449.
(W. II.).

597.
[Anon.], 'Philosophy of Apparitions', *Quarterly Review*, 48 (1832), 287–320.
(W. I.).

598.
[Anon.], *To the Electors of Roxburghshire, against the Candidature of Lord John Russell* [as Member of Parliament]. (Edinburgh, 1832).
(N. L. S.).

599–617.
Letters on Natural Magic Addressed to Sir Walter Scott, Bart. (London, 1832).
—1st American edition (New York, 1832).
—2nd edition (London, 1833).
—3rd edition (London, 1834).
—Another edition (New York, 1835).
—Another edition (New York, 1836).
—4th edition (London, 1838).
—Another edition (New York, 1838).
—Another edition (New York, 1839).
—5th edition (London, 1842).
—Another edition (New York, 1842).
—Another edition (New York, 1845).
—Another edition (New York, 1855).
—Another edition (London, 1856).
—New edition, with introductory chapters on the being and faculties of man and additional phenomena of natural magic, by J. A. Smith (London, 1868).
—Another edition (New York, 1870).
—New edition (London, 1882).
—Another edition (London, 1883).
—*Nouveau manuel de magie naturelle et amusante* ... (Paris, 1839).
(B. M. Cat., Nat. Un. Cat., Morse).

618.
D. B., 'Intelligence and Miscellaneous Articles: Singular Fog-Bow seen above Old Melrose', *London and Edinburgh Philosophical Magazine and Journal of Science*, 3rd series, *2* (1833), 151–152.

619–620.
'Observations on the Action of Light upon the Retina', *London and Edinburgh Philosophical Magazine and Journal of Science*, 3rd series, *2* (1833), 168–175; *Annalen der Physik und Chemie; von J. C. Poggendorff, 29* (1833), 339–352.
(R. S. C. 140).

621–622.
'Observations on the Absorption of Specific Rays, in Reference to the Undulatory Theory of Light', *London and Edinburgh Philosophical Magazine and Journal of Science*, 3rd series, *2* (1833), 360–363; *Annalen der Physik und Chemie; von J. C. Poggendorff, 28* (1833), 380–385.
(R. S. C. 141).

623–625.
'Notice Respecting Certain Changes of Colour in the Choroid Coat of the Eye of Animals', *London and Edinburgh Philosophical Magazine and Journal of Science*, 3rd series, *3* (1833), 288–289; *Notizen aus dem Gebiete der Natur und Heilkunde; von L. von Froriep, 38* (1833), 339–340; *Annalen der Physik und Chemie; von J. C. Poggendorff, 29* (1833), 479–480.
(R. S. C. 142).

626–629.
'On the Anatomical and Optical Structure of the Crystalline Lenses of Animals, particularly of the Cod', *Philosophical Transactions of the Royal Society of London, 123* (1833), 323–332; *ibid., 126* (1836), 35–48; *London and Edinburgh Philosophical Magazine and Journal of Science*, 3rd series, *8* (1836), 193–202; *Notizen aus dem Gebiete der Natur und Heilkunde; von L. von Froriep, 39* (1834), 152–154.
(R. S. C. 143).

630–635.
'Observations Relative to the Structure and Origin of the Diamond', *London and Edinburgh Philosophical Magazine and Journal of Science*, 3rd series, *3* (1833), 219–220; *ibid., 7* (1835), 245–250; *Proceedings of the Geological Society of London, 1* (1834), 446; *Transactions of the Geological Society of London*, new series, *3* (1835), 455–460; *Notizen aus dem Gebiete der Natur und Heilkunde; von L. von Froriep, 38* (1833), 213–215; *Annalen der Chemie und Pharmacie; von J. Liebig, Wöhler und Kopp, 18* (1836), 76–78.
(R. S. C. 151).

636.
'On Mineralogy', *Report of the British Association for the Advancement of Science ... 1831* (London, 1833), 60. (Signed article).

637.
'Description of an Instrument for Distinguishing Precious Stones and Minerals', *Report of the British Association for the Advancement of Science ... 1831* (London, 1833), 72–73. (Signed article).

638.
'On the Structure of the Crystalline Lens in Fishes, Birds, Reptiles and Quadrupeds', *Report of the British Association for the Advancement of Science ... 1831* (London, 1833), 81–82.
(R. S. C. 129).

639.
'On a New Analysis of Solar Light', *Report of the British Association for the Advancement of Science ... 1831* (London, 1833), 89–90. (Signed article).

640.
'Report on the Recent Progress of Optics', *Report of the British Association for the Advancement of Science ... 1832* (London, 1833), 308–322.
(R. S. C. 130).

641.
'On the Colour of Natural Bodies', *Report of the British Association for the Advancement of Science ... 1832* (London, 1833), 547–548. (Signed article).

642–645.
'On the Undulations Excited in the Retina by the Action of Luminous Points and Lines', *Report of the British Association for the Advancement of Science ... 1832* (London, 1833), 548–551; *Notizen aus dem Gebiete der Natur und Heilkunde; von L. von Froriep, 36* (1833), 241–246; *London and Edinburgh Philosophical Magazine and Journal of Science*, 3rd series, *1* (1832), 169–172; *Annalen der Physik und Chemie; von J. C. Poggendorff, 27* (1833), 490–496.
(R. S. C. 131).

646.
[Anon.], 'Life and Correspondence of Sir James Edward Smith', *Edinburgh Review, 57* (1833), 39–69.
(W. I.).

647–648.
[Anon.], 'British Lighthouse System', *Edinburgh Review, 57* (1833), 169–195; *ibid., 61* (1835), 528–531.
(W. I.).

649.
[Anon.], *Letter to the Right Hon. the Lord Provost.* An application for the Chair of Natural Philosophy in the University of Edinburgh with letters, additional certificates and testimonials in favour of Sir David Brewster (Edinburgh, 1833).
(B. M. Cat.).

650–653.
'Observations on the Lines of the Solar Spectrum, and on those produced by the Earth's Atmosphere, and by the Action of Nitrous Gas', *Transactions of the Royal Society of Edinburgh, 12* (1834), 519–530; *London and Edinburgh Philosophical Magazine and Journal of Science*, 3rd series, *8* (1836), 384–392; *Annalen der Physik und Chemie; von J. C. Poggendorff, 28* (1833), 50–64; *ibid., 33* (1834), 233–235.
(R. S. C. 144).

654–658.
'On the Colours of Natural Bodies', *Transactions of the Royal Society of Edinburgh, 12* (1834), 538–545; *London and Edinburgh Philosophical Magazine and Journal of Science*, 3rd series, *8* (1836), 468–474;

Notizen aus dem Gebiete der Natur und Heilkunde; von L. von Froriep, 49 (1836), 99–100, 129–135; *Annalen der Physik und Chemie; von J. C. Poggendorff,* 39 (1836), 476–484.
(R. S. C. 145).

659–660.
'Observations on the Supposed Vision of the Blood-Vessels of the Eye', *London and Edinburgh Philosophical Magazine and Journal of Science,* 3rd series, 4 (1834), 115–120; *Notizen aus dem Gebiete der Natur und Heilkunde von L. von Froriep,* 40 (1834), 81–87.
(R. S. C. 146).

661.
'On the Influence of Successive Impulses of Light upon the Retina', *London and Edinburgh Philosophical Magazine and Journal of Science,* 3rd series, 4 (1834), 241–245.
(R. S. C. 147).

662.
'Account of a Rhombohedral Crystallisation of Ice', *London and Edinburgh Philosophical Magazine and Journal of Science,* 3rd series, 4 (1834), 245–246. (Signed article).

663–664.
'Account of Two Experiments on Accidental Colours; with Observations on Their Theory', *London and Edinburgh Philosophical Magazine and Journal of Science,* 3rd series, 4 (1834), 353–354; *Notizen aus dem Gebiete der Natur und Heilkunde; von L. von Froriep,* 40 (1834), 325–327.
(R. S. C. 148).

665.
[Anon.], Review of the Bridgewater Bequest: Whewell's 'Astronomy and General Physics', *Edinburgh Review,* 58 (1834), 422–457.
(W. I).

666.
[Anon.], Review of Mrs. Somerville on 'The Physical Sciences', *Edinburgh Review,* 59 (1834), 154–171.
(W. I.).

667.
[Anon.], Review of Godwin's 'Lives of the Necromancers', *Edinburgh Review,* 60 (1834), 37–54.
(W. I.).

668.
[Anon.], Review of Dr. Roget's Bridgewater Treatise—'Animal and Vegetable Physiology', *Edinburgh Review,* 60 (1834), 142–179.
(W. I.).

669.
'Notice of the Optical Properties of a New Mineral supposed to be a Variety of Cymophane', *London and Edinburgh Philosophical Magazine and Journal of Science,* 3rd series, 6 (1835), 133–134. (Signed article).

670.
'Observations on the supposed Achromatism of the Eye', *London and Edinburgh Philosophical Magazine and Journal of Science,* 3rd series, 6 (1835), 161–164.
(R. S. C. 152).

671–673.
'On Certain Peculiarities in the Double Refraction and Absorption of Light Exhibited in the Oxalate of Chromium and Potash', *Philosophical Transactions of the Royal Society of London,* 125 (1835), 91–94; *London and Edinburgh Philosophical Magazine and Journal of Science,* 3rd series, 7 (1835), 436–439; *Annalen der Physik und Chemie; von J. C. Poggendorff,* 37 (1836), 315–318.
(R. S. C. 153).

674–675.
'Notice Respecting a Remarkable Specimen of Amber', *Report of the British Association for the Advancement of Science . . . 1834* (London, 1835), 574–575; *Journal für praktische Chemie; von Otto Linné Erdmann,* 6 (1835), 96–97.
(R. S. C. 149).

676.
'Remarks on the Value of Optical Characters in the Discrimination of Mineral Species', *Report of the British Association for the Advancement of Science . . . 1834* (London, 1835), 575.
(R. S. C. 150).

677.
[Anon.], 'The British Scientific Association', *Edinburgh Review,* 60 (1835), 363–394.
(W. I.).

678.
[Anon.], 'On the Universality of Goethe's Genius', *Fraser's Magazine,* 11 (1835), 11–12.
(W. II.).

679–680.
[Anon.], 'Parliamentary Report on Lighthouses', *Edinburgh Magazine,* 61 (1835), 221–241; Note, signed by Alan Stevenson, *ibid.,* 526–528; Reply by [? Brewster], *ibid.,* 528–531.
(W. I.).

681.
[Anon.], 'Ross's Voyage to the Arctic Regions', *Edinburgh Review,* 61 (1835), 417–453.
(W. I.).

682.
'Observations Relative to the Preceding Paper' [J. Johnston on Optical properties observed by David Brewster on Crystals of Chabasie] *London and Edinburgh Philosophical Magazine and Journal of Science,* 3rd series, 9 (1836), 170. (Signed article).

683.
'On the Action of Crystallised Surfaces upon Common and Polarised Light', *Report of the British Association for the Advancement of Science . . . 1836* (London, 1837), Part II, 13–16.
(R. S. C. 154).

684–687.
'On a Singular Development of Polarising Structure in the Crystalline Lens after Death', *Report of the British Association for the Advancement of Science . . . 1836* (London, 1837), Part II, 16–18; *Philosophical Transactions of the Royal Society of London,* 127 (1837), 253–258; *London and Edinburgh Philosophical Magazine and Journal of Science,* 3rd series, 12 (1838), 22–27; *Notizen aus dem Gebiete der Natur und Heilkunde; von L. von Froriep,* 6 (1838), 145–149.
(R. S. C. 155).

688.
'On the Cause, Prevention and Cure of Cataract', *Report of the British Association for the Advancement of Science . . . 1836* (London, 1837), Part II, 111–112. (Signed article).

689.
[Anon.], Review of 'Life and Works of Baron Cuvier', *Edinburgh Review,* 62 (1836), 265–296.
(W. I.).

690.
[Anon.], Review of 'Memoirs of Sir Humphrey Davy', *Edinburgh Review,* 63 (1836), 101–135.
(W. I.).

691–692.
'On the Connexion between the Phenomena of the Absorption of Light and the Colours of Thin Plates', *Philosophical Transactions of the Royal Society of London,* 127 (1837), 245–252; *London, Edinburgh and Dublin Philosophical Magazine and Journal of Science,* 3rd series, 21 (1842), 208–217.
(R. S. C. 156).

693.
[Anon.], Review of Lord Brougham's 'Discourse on Natural Theology', *Edinburgh Review*, 64 (1837), 263–302.
(W. I.).

694.
[Anon.], Review of Dr. Buckland's Bridgewater Treatise—'Geology and Mineralogy', *Edinburgh Review*, 65 (1837), 1–39.
(W. I.).

695.
[Anon.], Review of Dr. Bradley's 'Works and Correspondence', *Edinburgh Review*, 65 (1837), 119–132.
(W. I.).

696.
[Anon.], Review of Whewell's 'History of the Inductive Sciences', *Edinburgh Review*, 66 (1837), 110–151.
(W. I.).

697–698.
A Treatise on Magnetism, forming the article under that head in the seventh edition of the Encyclopaedia Britannica. (Edinburgh, 1837).
—Another edition (Edinburgh, 1838).
(B. M. Cat., Nat. Un. Cat.).

699–700.
A Treatise on the Microscope, forming the article under that head in the seventh edition of the Encyclopaedia Britannica. (Edinburgh, 1837).
—Another edition (Edinburgh, 1857).
(B. M. Cat., Nat. Un. Cat.).

701.
'On the Cause of the Optical Phenomena which take place in the Crystalline Lens during the Absorption of Distilled Water', *Report of the British Association for the Advancement of Science . . . 1837* (London, 1838), Part II, 11–12.
(R. S. C. 157).

702.
'On a New Property of Light', *Report of the British Association for the Advancement of Science . . . 1837* (London, 1838), Part II, 12–13.
(R. S. C. 158).

703.
'Notice of a New Structure in the Diamond', *Report of the British Association for the Advancement of Science . . . 1837* (London, 1838), Part II, 13–15.
(R. S. C. 159).

704.
'Analyse des Gmelinits oder Hydroliths', *Annalen der Chemie und Pharmacie; von J. Liebig, Wöhler, und Kopp*, 28 (1838), 342.
(R. S. C. 167).

705–706.
'On the Colours of Mixed Plates', *Philosophical Transactions of the Royal Society of London*, 128 (1838), 73–78; *London and Edinburgh Philosophical Magazine and Journal of Science*, 3rd series, 14 (1839), 191–196.
(R. S. C. 168).

707.
[Anon.], 'Weather Almanacks—the Late Frost', *The Monthly Chronicle*, 1 (1838), 76–84.
(W. III).

708.
[Anon.], 'Are the Planets Inhabited?' *The Monthly Chronicle*, 1 (1838), 101–115.
(W. III.).

709.
[Anon.], 'International Law of Copyright: Science', *The Monthly Chronicle*, 1 (1838), 163–166.
(W. III).

710.
[Anon.], Review of M. Comte's 'Course of Positive Philosophy', *Edinburgh Review*, 67 (1838), 271–308.
(W. I.).

711.
[Anon.], 'Fortifications of the Selenites', [moon dwellers] *The Monthly Chronicle*, 2 (1838), 150–154.
(W. III.).

712.
[Anon.], 'Great Discoveries in Astronomy', *The Monthly Chronicle*, 2 (1838), 517–524.
(W. III.).

713.
'On an Ocular Parallax in Vision, and on the Law of Visible Direction', *Report of the British Association for the Advancement of Science . . . 1838* (London, 1839), Part II, 7–9.
(R. S. C. 160).

714–715.
'On a New Phenomena of Colour in Certain Specimens of Fluor Spar', *Report of the British Association for the Advancement of Science . . . 1838* (London, 1839), Part II, 10–12; *Annales de Chimie, ou Recueil de Mémoires concernant la Chimie et les Arts qui en dépendent*, new series, 38 (1853), 376–377.
(R. S. C. 161).

716.
'An Account of Certain New Phenomena of Diffraction', *Report of the British Association for the Advancement of Science . . . 1838* (London, 1839), Part II, 12.
(R. S. C. 162).

717.
'On the Combined Action of Grooved Metallic and Transparent Surfaces upon Light', *Report of the British Association for the Advancement of Science . . . 1838* (London, 1839), Part II, 13.
(R. S. C. 163).

718–719.
'On a New Kind of Polarity in Homogeneous Light', *Report of the British Association for the Advancement of Science . . . 1838* (London, 1839), Part II, 13–14; *Annalen der Physik und Chemie; von J. C. Poggendorff*, 46 (1839), 481–484.
(R. S. C. 164).

720.
'On some Preparations of the Eye by Mr. Clay Wallace of New York', *Report of the British Association for the Advancement of Science . . . 1838* (London, 1839), Part II, 14–15.
(R. S. C. 165).

721.
'On the Structure of the Fossil Teeth of the Sauroid Fishes', *Report of the British Association for the Advancement of Science . . . 1838* (London, 1839), Part II, 90–91.
(R. S. C. 166).

722.
'Observations on Professor Plateau's Defence of his Theory of Accidental Colours', *London and Edinburgh Philosophical Magazine and Journal of Science*, 3rd series, 15 (1839), 435–441.
(R. S. C. 170).

723.
[Anon.], 'Statistics and Philosophy of Storms', *Edinburgh Review*, 68 (1839), 406–432.
(W. I.).

724.
[Anon.], 'The Sciences connected with Natural Theology', *The Monthly Chronicle*, 3 (1839), 97–115. [Review of Paley's *Natural Theology* edited with extensive notes and commentary by Lord Brougham].
(W. III.).

725.
[Anon.], Review of 'Life and Works of Thomas Telford', *Edinburgh Review*, 70 (1839), 1–47.
(W. I.).

726.
'Report respecting the Two Series of Hourly Meteorological Observations kept in Scotland', *Report of the British Association for the Advancement of Science . . . 1839* (London, 1840), 27–29.
(R. S. C. 169).

727–729.
'On the Optical Figures produced by the Disintegrated Surfaces of Crystals', *Transactions of the Royal Society of Edinburgh*, 14 (1840), 164–175; *Annalen der Chemie und Pharmacie; von J. Liebig, Wöhler, und Kopp*, 40 (1841) 127–131; *London, Edinburgh and Dublin Philosophical Magazine and Journal of Science*, 4th series, 5 (1853), 16–28.
(R. S. C. 180).

730–731.
[Anon.], 'Life and Discoveries of James Watt', *Edinburgh Review*, 70 (1840), 466–502; *Littell's Museum of Foreign Literature*, 41 (1846), 179–199.
(W. I., Poole).

732.
[Anon.], 'Scrope on Deer-stalking', *Edinburgh Review*, 71 (1840), 98–120.
(W. I.).

733.
[Anon.], Review of Goethe's 'Theory of Colours', *Edinburgh Review*, 72 (1840), 99–131.
(W. I.).

734.
'On Professor Powell's Measures of the Indices of Refraction for the Lines G and H in the Spectrum', *Report of the British Association for the Advancement of Science . . . 1840* (London, 1841), Part II, 5.
(R. S. C. 171).

735–736.
'On the Decomposition of Glass', *Report of the British Association for the Advancement of Science . . . 1840* (London, 1841), Part II, 5–6; *Annals of Electricity, Magnetism and Chemistry, and Guardian of Experimental Science; by William Sturgeon*, 5 (1840), 475–477.
(R. S. C. 172).

737–738.
'On the Rings of Polarised Light produced in Specimens of Decomposed Glass', *Report of the British Association for the Advancement of Science . . . 1840* (London, 1841), Part II, 6–7; *Annals of Electricity, Magnetism and Chemistry and Guardian of Experimental Science; by William Sturgeon*, 5 (1840), 477–478.
(R. S. C. 173).

739–740.
'On the Cause of the Increase of Colour by the Inversion of the Head', *Report of the British Association for the Advancement of Science . . . 1840* (London, 1841), Part II, 7–8; *Annals of Electricity, Magnetism and Chemistry, and Guardian of Experimental Science; by William Sturgeon*, 6 (1841), 59–61; *Annalen der Physik und Chemie; von J. C. Poggendorff*, 54 (1841), 137–139.
(R. S. C. 174).

741.
'On the Phenomena and Cause of Muscae Volitantes', *Report of the British Association for the Advancement of Science . . . 1840* (London, 1841), Part II, 8–9.
(R. S. C. 175).

742.
'A Brief Account of the Camera Obscura, and other Apparatus, used in making Daguerreotype Drawings', *Report of the British Association for the Advancement of Science . . . 1840* (London, 1841), Part II, 9. (Signed article).

743.
'On a Method of Illuminating Microscopic Objects', *Report of the British Association for the Advancement of Science . . . 1840* (London, 1841), Part II, 9–10; *Annals of Electricity, Magnetism, and Chemistry, and Guardian of Experimental Science; by William Sturgeon*, 6 (1841), 61–62.
(R. S. C. 176).

744.
'On an Improvement in the Polarising Microscope', *Report of the British Association for the Advancement of Science . . . 1840* (London, 1841), Part II, 10.
(R. S. C. 177).

745.
'Extract of a Letter from Col. Reid to Sir David Brewster on the Appearance of the Sun at Bermuda', *Report of the British Association for the Advancement of Science . . . 1840* (London, 1841), Part II, 10–11. (Signed article).

746.
'On a Remarkable Rainbow observed by Mr Bowman', *Report of the British Association for the Advancement of Science . . . 1840* (London, 1841), Part II, 12.
(R. S. C. 178).

747.
'Report respecting the Two Series of Hourly Meteorological Observations kept at Inverness and Kingussie from November 1st 1838 to November 1st 1839', *Report of the British Association for the Advancement of Science . . . 1840* (London, 1841), 349–352.
(R. S. C. 179).

748–750.
'On a Remarkable Property of the Diamond', *Philosophical Transactions of the Royal Society of London*, 131 (1841), 41–42; *Annalen der Physik und Chemie; von J. C. Poggendorff*, 58 (1843), 450–453; *London, Edinburgh and Dublin Philosophical Magazine and Journal of Science*, 4th series, 3 (1852), 284–286.
(R. S. C. 181).

751–754.
'On the Phenomena of Thin Plates of Solid and Fluid Substances exposed to Polarised Light', *Philosophical Transactions of the Royal Society of London*, 131 (1841), 43–58; *Annalen der Physik und Chemie; von J. C. Poggendorff*, 58 (1843), 453–456, 549–562; *London, Edinburgh and Dublin Philosophical Magazine and Journal of Science*, 3rd series, 32 (1848), 181–199.
(R. S. C. 182).

755–756.
'On the Compensations of Polarised Light, with the Description of a Polarimeter for Measuring Degrees of Polarisation', *Proceedings of the Royal Society of London*, 4 (1841), 306–307; *Transactions of the Irish Academy*, 19 (1843), 377–392.
(R. S. C. 183).

757.
'On the Colouring Matter of certain Land-shells from the Philippine Islands', *Proceedings of the Zoological Society of London*, 9 (1841), 15–16.
(R. S. C. 184).

758.
'Correction of an Error in Professor Dove's Letter on the Law of Storms', *London, Edinburgh and Dublin Philosophical Magazine and Journal of Science*, 3rd series, 18 (1841), 514–515. (Signed article).

759–775.
The Martyrs of Science; or, the lives of Galileo, Tycho Brahe, and Kepler. (London, 1841).
—1st American edition (New York, 1841).
—Another edition (New York, 1842).
—Another edition (New York, 1843).
—Another edition (New York, 1844).
—2nd edition (London, 1846).

—Another edition (New York, 1847).
—Another edition (New York, 1854).
—Another edition (New York, 1857).
—Another edition (New York, 1860).
—Another edition (New York, 1869).
—7th edition (London, 1870).
—Another edition (New York, 1872).
—A new edition with portraits (London, 1874).
—Another edition (New York, 1877).
—Another edition, (London, 1880).
—Another edition (London, 1903).
(B. M. Cat., Nat. Un. Cat.).

776.
'Report on The Hourly Observations at Inverness and Uist (*sic*) [Unst.]', *Report of the British Association for the Advancement of Science . . . 1841* (London, 1842), 329. (Signed article).

777.
'Report on the Erection of an Osler's Anemometer at Inverness', *Report of the British Association for the Advancement of Science . . . 1841* (London, 1842), 329. (Signed article).

778.
'Report on the State of Inquiry into the Action of Gaseous and other Media on the Solar Spectrum', *Report of the British Association for the Advancement of Science . . . 1841* (London, 1842), 329–330. (Signed article).

779.
[Anon.], Review of Whewell's 'Philosophy of the Inductive Sciences', *Edinburgh Review*, 74 (1842), 265–306.
(W. I.).

780.
[Anon.], Review of 'The Encyclopaedia Britannica, seventh edition', *Quarterly Review*, 70 (1842), 44–72.
(W. I.).

781.
N. N. N., 'Dynamics' *Encyclopaedia Britannica*, 21 vols., 7th edition, (Edinburgh, 1827–1842) VIII, 344–388.
(List).

782.
N. N. N., 'Electricity' *Encyclopaedia Britannica*, 21 vols., 7th edition, (Edinburgh, 1827–1842) VIII, 565–663.
(List).

783.
N. N. N., 'Hydrodynamics' *Encyclopaedia Britannica*, 21 vols., 7th edition, (Edinburgh, 1827–1842), XII, 1–109.
(List).

784.
[Anon.], 'Kaleidoscope' *Encyclopaedia Britannica*, 21 vols., 7th edition, (Edinburgh, 1827–1842) XII, 662–669.
(List).

785.
N. N. N., 'Magnetism' *Encyclopaedia Britannica*, 21 vols., 7th edition, (Edinburgh, 1827–1842) XIII, 685–774.
(List).

786.
[Anon.], 'Mechanics' *Encyclopaedia Britannica*, 21 vols., 7th edition, (Edinburgh, 1827–1842) XIV, 344–459.
(List).

787.
[Anon.], 'Micrometer' *Encyclopaedia Britannica*, 21 vols., 7th edition, (Edinburgh, 1827–1842) XV, 10–27.
(List).

788.
N. N. N., 'Microscope' *Encyclopaedia Britannica*, 21 vols., 7th edition, (Edinburgh, 1827–1842) XV, 27–57.
(List).

789.
N. N. N., 'Newton, Sir Isaac' *Encyclopaedia Britannica*, 21 vols., 7th edition, (Edinburgh, 1827–1842) XVI, 175–181.
(List).

790.
[Anon.], 'Optics' *Encyclopaedia Britannica*, 21 vols., 7th edition, (Edinburgh, 1827–1842) XVI, 348–514.
(List).

791.
[Anon.], 'Voltaic Electricity' *Encyclopaedia Britannica*, 21 vols., 7th edition, (Edinburgh, 1827–1842) XX, 664–700.
(List).

792–793.
'On a New Property of the Rays of the Spectrum, with Observations on the Explanation of it given by the Astronomer Royal, on the Principle of the Undulatory Theory', *Report of the British Association for the Advancement of Science . . . 1842* (London, 1843), Part II, 12; *Annali di Fisica, Chimica, etc; da G. Majocchi e F. Selmi*, 10 (1843), 261–265.
(R. S. C. 185).

794.
'On the Dichroism of the Palladio-chlorides of Potassium and Ammonium', *Report of the British Association for the Advancement of Science . . . 1842* (London, 1843), Part II, 13.
(R. S. C. 186).

795.
'On the Existence of a New Neutral Point and two Secondary Neutral Points', *Report of the British Association for the Advancement of Science . . . 1842* (London, 1843), Part II, 13.
(R. S. C. 187).

796–797.
'On Crystalline Reflexion', *Report of the British Association for the Advancement of Science . . . 1842* (London, 1843), Part II, 13–14; *Annals of Electricity, Magnetism and Chemistry, and Guardian of Experimental Science; by William Sturgeon*, 9 (1842), 458–462.
(R. S. C. 188).

798.
'On a very Curious Fact connected with Photography, discovered by Mr Möser of Königsberg', *Report of the British Association for the Advancement of Science . . . 1842* (London, 1843), Part II, 14.
(R. S. C. 189).

799.
'On the Dichroism of a Solution of Stramonium in Ether', *Report of the British Association for the Advancement of Science . . . 1842* (London, 1843), Part II, 14. (Signed article).

800.
'On the Geometric Forms and Laws of Illumination on the Spaces which receive the Solar Rays Transmitted through Quadrangular Apertures', *Report of the British Association for the Advancement of Science . . . 1842* (London, 1843), Part II, 15.
(R. S. C. 190).

801.
'On Luminous Lines in certain Flames corresponding to the Defective Lines in the Sun's Light', *Report of the British Association for the Advancement of Science . . . 1842* (London, 1843), Part II, 15.
(R. S. C. 191).

802.
'On the structure of a part of the Solar Spectrum hitherto unexamined', *Report of the British Association for the Advancement of Science . . . 1842* (London, 1843), Part II, 15.
(R. S. C. 192).

803.
'On the Luminous Bands in the Spectra of Various Flames', *Report of the British Association for the Advancement of Science . . . 1842* (London, 1843), Part II, 15–16.
(R. S. C. 193).

804.
'On the Erection of one of Mr Osler's Anemometers at Inverness, one of the stations at which Hourly Observations with the Barometer and Thermometer have been made, at the Request of the British Association', *Report of the British Association for the Advancement of Science . . . 1842* (London, 1843), 206. (Signed article).

805.
'On the Hourly Series of Meteorological Observations made at Inverness during the Meteorological Year from 1st November 1840 to 1st November 1841', *Report of the British Association for the Advancement of Science . . . 1842* (London, 1843), 206. (Signed article).

806–807.
'On the Cause of the Colours in Irridescent Agate', *London, Edinburgh and Dublin Philosophical Magazine and Journal of Science*, 3rd series, *22* (1843), 213–215; *Annalen der Physik und Chemie; von J. C. Pogendorff*, *61* (1844), 134–138.
(R. S. C. 196).

808–809.
'On the Combination of Prolonged Direct Luminous Impressions on the Retina, with their Complementary Impressions', *London, Edinburgh and Dublin Philosophical Magazine and Journal of Science*, 3rd series, *22* (1843), 434–435; *Annalen der Physik und Chemie; von J. C. Poggendorff*, *61* (1844), 138–140.
(R. S. C. 197).

810.
[Anon.], 'Photogenic Drawing, or Drawing by the Agency of Light', *Edinburgh Review*, *76* (1843), 309–344.
(W. I.).

811.
[Anon.], Review of Wilson's 'Voyage round Scotland and the Isles', *Edinburgh Review*, *77* (1843), 170–190.
(W. I.).

812.
[Anon.], Review of William Scrope's 'Days and Nights of Salmon Fishing', *Edinburgh Review*, *78* (1843), 87–113.
(W. I.).

813.
[Anon.], 'Hay on Harmonious Colouring and Form', *Edinburgh Review*, *78* (1843), 300–326.
(W. I.).

814.
'Meteorological Observations at Inverness', *Report of the British Association for the Advancement of Science . . . 1843* (London, 1844), 292. (Signed article).

815.
'Meteorological Observations at Unst', *Report of the British Association for the Advancement of Science . . . 1843* (London, 1844), 293. (Signed article).

816.
'On the Action of Different Bodies on the Spectrum', *Report of the British Association for the Advancement of Science . . . 1843* (London, 1844), 293–294. (Signed article).

817.
'Notice of an Experiment on the Ordinary Refraction of Iceland spar', [with remarks by MacCullagh], *Report of the British Association for the Advancement of Science . . . 1843* (London, 1844), Part II, 7–8.
(R. S. C. 194).

818.
'On the Action of Two Blue Oils upon Light', *Report of the British Association for the Advancement of Science . . . 1843* (London, 1844), Part II, 8.
(R. S. C. 195).

818–821.
'On the Law of Visible Position in Single and Binocular Vision, and on the Representation of Solid Figures by the Union of Dissimilar Plane Pictures on the Retina', *Transactions of the Royal Society of Edinburgh*, *12* (1844), 349–368; *London, Edinburgh and Dublin Philosophical Magazine and Journal of Science*, 3rd series, *24* (1844), 356–365, 439–455.
(R. S. C. 204).

822–823.
'On the Optical Phenomena, Nature, and Locality of Muscae Volitantes, with Observations on the Structure of the Vitreous Humour, and on the Vision of Objects placed within the Eye', *Transactions of the Royal Society of Edinburgh*, *15* (1844), 377–386; *Notizen aus dem Gebiete der Natur und Heilkunde; von M. J. Schleiden und Rob. Froriep*, 3rd series, *6* (1848), 209–214.
(R. S. C. 205).

824–825.
'On the Conversion of Relief by Inverted Vision', *Transactions of the Royal Society of Edinburgh*, *15* (1844), 657–662; *London, Edinburgh and Dublin Philosophical Magazine and Journal of Science*, 3rd series, *30* (1847), 432–437.
(R. S. C. 206).

826–827.
'On the Knowledge of Distance given by Binocular Vision', *Transactions of the Royal Society of Edinburgh*, *15* (1844), 663–675; *London, Edinburgh and Dublin Philosophical Magazine and Journal of Science*, 3rd series, *30* (1847), 305–318.
(R. S. C. 207 and 222).

828.
'Observations on Colour Blindness, or Insensibility to the Impressions of Certain Colours', *London, Edinburgh and Dublin Philosophical Magazine and Journal of Science*, 3rd series, *25* (1844), 134–141.
(R. S. C. 208).

829.
'Sopra un fenomeno della visione', *Annali di Fisica, Chimica, etc.; da G. Majocchi e F. Selmi*, *16* (1844), 24–25.
(R. S. C. 303).

830.
[Anon.], Review of Flourens's 'Éloge Historique de Baron Cuvier', *North British Review*, *1* (1844), 1–41.
(W. I.).

831.
[Anon.], Review of Forbes's 'Travels through the Alps of Savoy', *Edinburgh Review*, *80* (1844), 135–163.
(W. I.).

832.
[Anon.], 'Pascal's Life, Writings, and Discoveries', *North British Review*, *1* (1844), 285–326.
(W. I.).

833.
[Anon.], 'Harris on Thunderstorms and Protection from Lightning', *Edinburgh Review*, *80* (1844), 444–473.
(W. I.).

834–836.
[Anon.], 'The Earl of Rosse's Reflecting Telescopes', *North British Review*, *2* (1844), 175–212; *Littell's Living Age*, *3* (1844), 404–419; *Eclectic Magazine*, *5* (1845), 49–71.
(W. I., Poole).

837.
'On the Hourly Meteorological Observations carried out at Inverness at the Expense of the British Association, by Mr Thomas Mackenzie; a Provisional Report', *Report of the British Association for the Advancement of Science . . . 1844* (London, 1845), 391. (Signed article).

838.
'A Notice explaining the Cause of an Optical Phenomena observed by the Rev. W. Selwyn', *Report of the British Association for the Advancement of Science . . . 1844* (London, 1845), Part II, 8.
(R. S. C. 198).

839.
'An Account of the Cause of Colours in the precious Opal', *Report of the British Association for the Advancement of Science . . . 1844* (London, 1845), Part II, 9.
(R. S. C. 199).

840.
'A Notice Respecting the Cause of Beautiful White Rings which are seen around a Luminous Body when looked at through Certain Specimens of Calcareous Spar', *Report of the British Association for the Advancement of Science . . . 1844* (London, 1845), Part II, 9.
(R. S. C. 200).

841.
'On Crystals in the Cavities of Topaz, which are Dissolved by Heat and Re-crystallised on Cooling', *Report of the British Association for the Advancement of Science . . . 1844* (London, 1845), Part II, 9–10.
(R. S. C. 201).

842.
'On a Singular Effect of the Juxtaposition of Certain Colours under Particular Circumstances', *Report of the British Association for the Advancement of Science . . . 1844* (London, 1845), Part II, 10.
(Signed article).

843.
'On the Accommodation of the Eye to Various Distances', *Report of the British Association for the Advancement of Science . . . 1844* (London, 1845), Part II, 10–11.
(R. S. C. 202).

844.
'Account of a Series of Experiments on the Polarisation of Light by Rough Surfaces and White Dispersing Surfaces', *Report of the British Association for the Advancement of Science . . . 1844* (London, 1845), Part II, 11.
(R. S. C. 203).

845–846.
'On the Optical Properties of Greenockite', *Proceedings of the Royal Society of Edinburgh, 1* (1845), 342–343; *Annalen der Physik und Chemie; von J. C. Poggendorff, 58* (1843), 94–95.
(R. S. C. 213).

847.
'On Certain Negative Actions of Light', *Proceedings of the Royal Society of Edinburgh, 1* (1845), 422–424.
(R. S. C. 214).

848–849.
'Sur la Polarization de la Lumière Atmosphérique', *Comptes Rendus Hebdomadaires des Séances de l'Académie des Sciences, 20* (1845), 801–802; *Annalen der Physik und Chemie; von J. C. Poggendorff, 66* (1845), 456–457.
(R. S. C. 215).

850.
'Observations Connected with the Discovery of the Composition of Water', *London, Edinburgh and Dublin Philosophical Magazine and Journal of Science*, 3rd series, *27* (1845), 195–197. [Self-defence from an attack in *Quarterly Review* on an article by D. B. in the *Edinburgh Review*. Reply to this, *ibid.*, 446–450.]
(R. S. C. 216).

851.
[Anon.], 'Eusebe Salverte on the Occult Sciences', *North British Review, 3* (1845), 1–39.
(W. I.).

852–853.
[Anon.], Review of 'Vestiges of the Natural History of Creation', *North British Review, 3* (1845), 470–515; *Littell's Living Age, 6* (1845), 564–582.
(W. I., Poole).

854.
[Anon.], Review of Baron Humboldt's 'Kosmos, a General Survey of the Physical Phenomena of the Universe', *North British Review, 4* (1845), 202–254.
(W. I.).

855.
[Anon.], 'Magic', *Hogg's Weekly Instructor, 2* (1845–46), 65–67. ('. . . our obligations to an able article in . . . the *North British Review*'.)

856.
'On a New Polarity of Light with an Examination of Mr Airy's Explanation of it on the Undulatory Theory', *Report of the British Association for the Advancement of Science . . . 1845* (London, 1846), Part II, 7–8.
(R. S. C. 209).

857.
'Notice of Two New Properties of the Retina', *Report of the British Association for the Advancement of Science . . . 1845* (London, 1846), Part II, 8–9.
(R. S. C. 210).

858.
'On the Rotation of Minute Crystals in the Cavities of Topaz', *Report of the British Association for the Advancement of Science . . . 1845* (London, 1846), Part II, 9. (Signed article).

859.
'On the Condition of Topaz subsequent to the Formation of Certain Cavities within it', *Report of the British Association for the Advancement of Science . . . 1845* (London, 1846), Part II, 9. (Signed article).

860.
'An Improvement in the Method of taking Positive Talbotypes', *Report of the British Association for the Advancement of Science . . . 1845* (London, 1846), Part II, 10–11.
(R. S. C. 211).

861.
'On Fog-Rings observed in America', *Report of the British Association for the Advancement of Science . . . 1845* (London, 1846), Part II, 19.
(R. S. C. 212).

862.
'On the Law of Daily Temperature', *London, Edinburgh and Dublin Philosophical Magazine and Journal of Science*, 3rd series, *29* (1846), 341–353. (Signed article).

863.
[Anon.], Review of Arago's 'Éloge Historique de Joseph Fourier', *North British Review, 4* (1846), 380–412.
(W. I.).

864.
[Anon.], Review of 'Explanations', by the Author of the 'Vestiges of the Natural History of Creation', *North British Review, 4* (1846), 487–504.
(W. I.).

865.
[Anon.], Review of Sir R. Murchison, M. de Verneuil, and Count Keyserling's 'The Geology of Russia in Europe', *North British Review, 5* (1846), 178–221.
(W. I.).

866.
[Anon.], Review of Baron Humboldt's 'Researches in Central Asia', *North British Review*, 5 (1846), 454–503.
(W. I.).

867.
[Anon.], 'Captain Smith and Dr Nichol on Celestial Objects', *North British Review*, 6 (1846), 206–255.
(W. I.).

868–870.
'Notice of a New Property of Light Exhibited in the Action of Chrysammate of Potash upon Common and Polarised light', *Report of the British Association for the Advancement of Science . . . 1846* (London, 1847), Part II, 7–8; *Annalen der Physik und Chemie; von J. C. Poggendorff*, 69 (1846), 552–554; *London, Edinburgh and Dublin Philosophical Magazine and Journal of Science*, 3rd series, 29 (1846), 331–332.
(R. S. C. 217).

871–872.
'Reply to the Astronomer Royal on the New Analysis of Solar Light', *London, Edinburgh and Dublin Philosophical Magazine and Journal of Science*, 3rd series, 30 (1847), 153–158; *Annalen der Physik und Chemie; von J. C. Poggendorff*, 71 (1847), 397–405.
(R. S. C. 221).

873–874.
'Observation on the Analysis of the Spectrum by Absorption', *London, Edinburgh and Dublin Philosophical Magazine and Journal of Science*, 3rd series, 30 (1847), 461–462; *Annalen der Physik und Chemie; von J. C. Poggendorff*, 75 (1848), 81–88.
(R. S. C. 223).

875–876.
'On the Modification of the Doubly Refracting and Physical Structure of Topaz, by Elastic Forces Emanating from Minute Cavities', *London, Edinburgh and Dublin Philosophical Magazine and Journal of Science*, 3rd series, 31 (1847), 101–104; *Transactions of the Royal Society of Edinburgh*, 16 (1849), 7–10.
(R. S. C. 224).

877.
'On the Polarisation of the Atmosphere', *London, Edinburgh and Dublin Philosophical Magazine and Journal of Science*, 3rd series, 31 (1847), 444–454.
(R. S. C. 225).

878–879.
'On the Existence of Crystals with Different Primitive Forms and Physical Properties in the Cavities of Minerals; with Additional Observations on the New Fluids in which they occur', *London, Edinburgh and Dublin Philosophical Magazine and Journal of Science*, 3rd series, 31 (1847), 497–510; *Transactions of the Royal Society of Edinburgh*, 16 (1879), 11–22.
(R. S. C. 226 and 241).

880.
[Anon.], 'Watt and Cavendish—Controversy Respecting the Composition of Water', *North British Review*, 6 (1847), 473–508.
(W. I.).

881.
[Anon.], 'Mr Adams' and M. Le Verrier's Researches, respecting the New Planet Neptune', *North British Review*, 7 (1847), 207–246.
(W. I.).

882.
[Anon.], 'Photography', *North British Review*, 7 (1847), 465–504.
(W. I.).

883.
[Anon.], Review of Sir J. C. Ross 'Antarctic Voyage of Discovery', and Lieut. Wilkes 'Exploring Expedition', *North British Review*, 8 (1847), 177–217.
(W. I.).

884–885.
[Anon.], 'Goring, Pritchard, Chevalier and Fischer, on the Construction and Use of Microscopes', *North British Review*, 8 (1847), 258–264; *Littell's Living Age*, 16 (1847), 227–231.
(W. I., Poole).

886.
'On the Polarisation of the Atmosphere', *Report of the British Association for the Advancement of Science . . . 1847* (London, 1848), Part II, 32. (Signed article).

887.
'On a New Species of Polarisation Related to the Direction of the Grooves in Grooved Surfaces', *Report of the British Association for the Advancement of Science . . .1847* (London, 1848), Part II, 32. (Signed article).

888.
'On the Diffraction Bands produced by the Edges of Thin Plates, whether Solid or Fluid', *Report of the British Association for the Advancement of Science . . . 1847* (London, 1848), Part II, 33.
(R. S. C. 218).

889.
'On the Dark Lines in the Portion of the Red Space Beyond the Red Extremity of the Spectrum as seen by Fraunhöfer', *Report of the British Association for the Advancement of Science . . . 1847* (London, 1848), Part II, 33.
(R. S. C. 219).

890.
'An Account of the Functions of the Membranes of the Eye at the Foramen Centrale of Soemmering', *Report of the British Association for the Advancement of Science . . . 1847* (London, 1848), Part II, 33.
(R. S. C. 220).

891.
'On the Conversion of Relief in a Drawing, by Inverting the Drawing and Viewing it with a Lens of Short Focus', *Report of the British Association for the Advancement of Science . . . 1847* (London, 1848), Part II, 33. (Signed article).

892–893.
'On the Optical Phenomena, Nature, and Locality of Muscae Volitantes; with Observations on the Structure of the Vitreous Humour, and on the Vision of Objects placed within the Eye', *London, Edinburgh and Dublin Philosophical Magazine and Journal of Science*, 3rd series, 32 (1848) 1–11; *Transactions of the Royal Society of Edinburgh*, 15 (1848) 377–385.
(R. S. C. 232).

894.
'On the Distinctness of Vision Produced in Certain Cases by the use of the Polarising Apparatus in Microscopes', *London, Edinburgh and Dublin Philosophical Magazine and Journal of Science*, 3rd series, 32 (1848), 161–165.
(R. S. C. 233).

895–898.
'On the Decomposition of Light within Solid and Fluid Bodies', *London, Edinburgh and Dublin Philosophical Magazine and Journal of Science*, 3rd series, 32 (1848), 401–412; *Transactions of the Royal Society of Edinburgh*, 16 (1849), 111–122; *Supplément à la Bibliothèque Universelle et Revue Suisse. Archives des Sciences Physiques et Naturelles*, 9 (1848), 118–132; *Annalen der Physik und Chemie; von J. C. Poggendorff*, 73 (1848), 531–548.
(R. S. C. 231 and 242).

899.
'Observations on the Elementary Colours of the Spectrum, in reply to M. Melloni', *London, Edinburgh and Dublin Philosophical Magazine and Journal of Science*, 3rd series, 32 (1848), 489–494.
(R. S. C. 234).

900.
'On the phenomena of Luminous Rings in Calcareous Spar and

Beryl, as produced by Tubular Cavities containing the Two New Fluids', *London, Edinburgh and Dublin Philosophical Magazine and Journal of Science*, 3rd series, *33* (1848), 489–493.
(R. S. C. 235).

901–902.
[Anon.], Review of Sir John Herschel's 'Astronomical Observations', *North British Review, 8* (1848), 491–533; *Littell's Living Age, 16* (1848), 577–596.
(W. I., Poole).

903.
[Anon.], Review of Mrs. Somerville's 'Physical Geography', *North British Review, 9* (1848), 141–186.
(W. I.).

904.
[Anon.], Review of Johnston's 'Physical Atlas', *North British Review, 9* (1848), 359–369.
(W. I.).

905–906.
[Anon.], Review of Mr. Brooke's 'Journals of a Residence in Borneo', *North British Review, 9* (1848), 432–471; *Eclectic Magazine, 15* (1848), 228–250.
(W. I., Poole).

907–908.
[Anon.], Review of Mr. Britton's 'Authorship of Junius Elucidated', *North British Review, 10* (1848), 97–143; *Eclectic Magazine, 16* (1849), 160–186.
(W. I., Poole).

909.
[Anon.], 'Von Humboldt', *Hogg's Instructor, 1* (1848), 273–276.
(Quotes Brewster; *cf.* 854 and 866).

910.
'On the Compensation of Impressions, moving over the Retina, as seen in Railway Travelling', *Report of the British Association for the Advancement of Science . . . 1848* (London, 1849), Part II, 47–48.
(R. S. C. 227).

911.
'On the Vision of Distance as given by Colour', *Report of the British Association for the Advancement of Science . . . 1848* (London, 1849), Part II, 48.
(R. S. C. 228).

912.
'On the Visual Impressions upon the Foramen Centrale of the Retina', *Report of the British Association for the Advancement of Science . . . 1848* (London, 1849), Part II, 48–49.
(R. S. C. 229).

913.
'An Examination of Bishop Berkeley's "New Theory of Vision"', *Report of the British Association for the Advancement of Science . . . 1848* (London, 1849), Part II, 49.
(R. S. C. 230).

914–916.
[Anon.], Review of Macaulay's 'History of England', *North British Review, 10* (1849), 367–424; *Eclectic Magazine, 16* (1849), 522–553; *Littell's Living Age, 21* (1849), 49–75.
(W. I., Poole).

917.
[Anon.], Review of Layard's 'Nineveh and Its Remains', *North British Review, 11* (1849), 209–253.
(W. I.).

918.
[Anon.], 'The Railway System of Great Britain', *North British Review, 11* (1849), 569–617.
(W. I.).

919–921.
[Anon.], Review of Humboldt's 'Aspects of Nature in Different Lands', *North British Review, 12* (1849), 225–264; *Littell's Living Age, 23* (1849), 587–605; *Eclectic Magazine, 19* (1850), 374–396.
(W. I., Poole).

922.
[Anon.], 'Sir David Brewster', *Hogg's Instructor, 3* (1849), 289–291.
(Dictated by Brewster ?).

923.
'Description of a Binocular Camera', *Report of the British Association for the Advancement of Science . . . 1849* (London, 1850), Part II, 5.
(R. S. C. 236).

924.
'Improvement of the Photographic Camera', *Report of the British Association for the Advancement of Science . . . 1849* (London, 1850), Part II, 5.
(R. S. C. 237).

925.
'On a New Form of Lenses, and their Application to the Construction of Two Telescopes or Microscopes of Exactly Equal Optical Power', *Report of the British Association for the Advancement of Science . . . 1849* (London, 1850), Part II, 6.
(R. S. C. 238).

926.
'Notice of Experiments on Circular Crystals', *Reports of the British Association for the Advancement of Science . . . 1849* (London, 1850), Part II, 6.
(R. S. C. 239).

927.
'Additional Observations on Berkeley's Theory of Vision', *Report of the British Association for the Advancement of Science . . . 1849* (London, 1850), Part II, 6. (Signed article).

928–929.
'An account of a New Stereoscope', *Report of the British Association for the Advancement of Science . . . 1849* (London, 1850), Part II, 6–7; *Edinburgh New Philosophical Journal, 48* (1850), 150–151.
(R. S. C. 240).

930.
'Sur quelques phénomènes de polarization qui ont rapport avec la diffraction opérée par les surfaces rayées', *Comptes Rendus Hebdomadaires des Séances de l'Académie des Sciences, 30* (1850), 496–498.
(R. S. C. 248).

931.
'Observations sur les points neutres de l'atmosphère découverts par MM. Arago et Babinet', *Comptes Rendus Hebdomadaires des Séances de l'Académie des Sciences, 30* (1850), 532–536.
(R. S. C. 249).

932–933.
'Observations sur le spectre solaire', *Comptes Rendus Hebdomadaires des Séances de l'Académie des Sciences, 30* (1850), 578–581; *Annalen der Physik und Chemie; von J. C. Poggendorff, 81* (1850), 471–476.
(R. S. C. 250).

934–935.
[Anon.], Review of Hugh Miller's 'Footprints of the Creator', *North British Review, 12* (1850), 443–481; *Littell's Living Age, 25* (1850), 145–162.
(W. I., Poole).

936–937.
[Anon.], Review of Hunt's 'Poetry of Science', *North British Review, 13* (1850), 117–158; *Eclectic Magazine, 20* (1850), 289–312.
(W. I., Poole).

938.
[Anon.], 'Messrs. Stephenson and Fairbairn's Tubular Bridges', *North British Review*, 13 (1850), 399–416.
(W. I.).

939–940.
[Anon.], 'British Association for the Advancement of Science', *North British Review*, 14 (1850), 235–287; *Eclectic Magazine*, 22 (1851), 145–169.
(W. I., Poole).

941.
[Anon.], 'Aspects of Nature', *Hogg's Instructor*, 4 (1850), 245–249.
(*Cf.* 909).

942.
'Presidential Address', *Report of the British Association for the Advancement of Science . . . 1850* (London, 1851), xxii–xliv.
(Morse).

943.
'Notice on the Artificial Magnets made by M. Logeman, by the Process of M. Elias', *Report of the British Association for the Advancement of Science . . . 1850* (London, 1851), Part II, 4.
(R. S. C. 243).

944.
'On a New Membrane investing the Crystalline Lens of the Ox', *Report of the British Association for the Advancement of Science . . . 1850* (London, 1851), Part II, 4–5.
(R. S. C. 244).

945.
'On the Optical Properties of the Cyanurets of Platinum and Magnesia, and of Barytes and Platinum', *Report of the British Association for the Advancement of Science . . . 1850* (London, 1851), Part II, 5.
(R. S. C. 245).

946.
'On the Polarising Structure of the Eye', *Report of the British Association for the Advancement of Science . . . 1850* (London, 1851), Part II, 5–6.
(R. S. C. 246).

947.
'On some new Phenomena in the Polarisation of the Atmosphere', *Report of the British Association for the Advancement of Science . . . 1850* (London, 1851), Part II, 6.
(R. S. C. 247).

948.
'Notice Regarding Recent Improvements in Photography', *Report of the British Association for the Advancement of Science . . . 1850* (London, 1851), Part II, 6. (Signed article).

949–952.
'Description of Several New and Simple Stereoscopes for Exhibiting, as Solids, one or more Representations of them on a Plane', *Transactions of the Royal Scottish Society of Arts*, 3 (1851), 247–258; *London, Edinburgh and Dublin Philosophical Magazine and Journal of Science*, 4th series, 3 (1852), 16–26.
Das Stereoskop von Dav. Brewster und seine neuesten Verbesserungen, Modificationen und Vervollkommnungen von Duboscq und Prof. Helmholtz [Translated from articles in the *Transactions of the Royal Scottish Society of Arts*]. (Quedlinburg, 1859).
—Another edition (Quedlinburg, 1908).
(B. M. Cat., R. S. C. 251).

953–954.
'Account of a Binocular Camera, and of a Method of Obtaining Drawings of full length and Colossal Statues, and of Living Bodies, which can be exhibited as Solids by the Stereoscopes', *Transactions of the Royal Scottish Society of Arts*, 3 (1851), 259–264; *London, Edinburgh and Dublin Philosophical Magazine and Journal of Science*, 4th series, 3 (1852), 26–31.
(R. S. C. 252).

955.
[Anon.], Review of Sir Charles Lyell's 'Travels in North America', *North British Review*, 14 (1851), 541–584.
(W. I.).

956.
[Anon.], Review of Arago's 'Life of Carnot', *North British Review*, 15 (1851), 185–227.
(W. I.).

957.
[Anon.], 'Mr Babbage on the Exposition of 1851', *North British Review*, 15 (1851), 529–568.
(W. I.).

958.
[Anon.], 'Prize Essays on the Peace Congress', *North British Review*, 16 (1851), 1–48.
(W. I.).

959.
[Anon.], 'Hugh Miller', *Hogg's Instructor*, 6 (1851), 81–84. (Long quote from 934).

960.
'Notice of a Chromatic Stereoscope', *London, Edinburgh and Dublin Philosophical Magazine and Journal of Science*, 4th series, 3 (1852), 31–32.
(R. S. C. 256).

961.
'Explanation of an Optical Illusion', *London, Edinburgh and Dublin Philosophical Magazine and Journal of Science*, 4th series, 3 (1852), 55–57.
(R. S. C. 257).

962.
'On the Development and Extinction of Regular Doubly Refracting Structures in the Crystalline Lenses of Animals after Death', *London, Edinburgh and Dublin Philosophical Magazine and Journal of Science*, 4th series, 3 (1852), 192–198.
(R. S. C. 258).

963.
Address delivered to Members of the Edinburgh Philosophical Institution, on the 11th November 1851 (London and Glasgow, 1852).
(B. M. Cat., family volume).

964.
[Anon.], 'The Peace Congress', *Hogg's Instructor*, new series, 8 (1852), 21–22. (Brewster was president of Peace Congress).

965.
[Anon.], 'Arctic Searching Expeditions', *North British Review*, 16 (1852), 445–491.
(W. I.).

966.
[Anon.], 'Binocular Vision and the Stereoscope', *North British Review*, 17 (1852), 165–204.
(W. I.).

967.
[Anon.], 'Prince Albert's Industrial College of Arts and Manufacturers', *North British Review*, 17 (1852), 519–558.
(W. I.).

968–969.
[Anon.], 'The Diamond—its History, Properties and Origin', *North British Review*, 18 (1852), 186–234; *Eclectic Magazine*, 28 (1852), 1–24.
(W. I., Poole).

970.
'Notice of a Tree struck by Lightning in Clandeboyne Park', *Report of the British Association for the Advancement of Science . . . 1852* (London, 1853), Part II, 2–3.
(R. S. C. 253).

971.
'Account of a Case of Vision without Retina', *Report of the British Association for the Advancement of Science ... 1852* (London, 1853), Part II, 3.
(R. S. C. 254).

972.
'On the Form of Images produced by Lenses and Mirrors of Different Sizes', *Report of the British Association for the Advancement of Science ... 1852* (London, 1853), Part II, 3–6.
(R. S. C. 255).

973.
'On Certain Phenomena of Diffraction', *Report of the British Association for the Advancement of Science ... 1852* (London, 1853), Part II, 24–25. (Signed article).

974.
'Observations on the Diamond', *Report of the British Association for the Advancement of Science ... 1852* (London, 1853), Part II, 41–42. (Signed article).

975–977.
'On the Optical Phenomena and Crystallisation of Tourmaline, Titanium, and Quartz, within Mica, Amethyst, and Topaz', *Transactions of the Royal Society of Edinburgh*, 20 (1853), 547–554; *Report of the British Association for the Advancement of Science ... 1853* (London, 1854), Part II, 3–4; *London, Edinburgh and Dublin Philosophical Magazine and Journal of Science*, 4th series, 6 (1853), 265–272.
(R. S. C. 259).

978–980.
'On the Production of Crystalline Structure in Crystallised Powders by Compression and Traction', *Transactions of the Royal Society of Edinburgh*, 20 (1853), 555–560; *Report of the British Association for the Advancement of Science ... 1853* (London, 1854), Part II, 3; *London, Edinburgh and Dublin Philosophical Magazine and Journal of Science*, 4th series, 6 (1853), 260–264.
(R. S. C. 260).

981.
'On Circular Crystals', *Transactions of the Royal Society of Edinburgh*, 20 (1853), 607–624.
(R. S. C. 261).

982–983.
'On the Cavities in Amber containing Gases and Fluids', *London, Edinburgh and Dublin Philosophical Magazine and Journal of Science*, 4th series, 5 (1853), 233–235; *Annalen der Physik und Chemie; von J. C. Poggendorff*, 91 (1854), 605–607.
(R. S. C. 262).

984–985.
'Account of a Remarkable Fluid Cavity in Topaz', *London, Edinburgh and Dublin Philosophical Magazine and Journal of Science*, 4th series, 5 (1853), 235–236; *Annalen der Physik und Chemie; von J. C. Poggendorff*, 91 (1854), 607–608.
(R. S. C. 263).

986.
[Anon.], 'The Universe and its Laws', *North British Review*, 18 (1853), 491–525.
(W. I.).

987.
[Anon.], Review of Layard's 'Discoveries in Nineveh and Babylon', *North British Review*, 19 (1853), 255–296.
(W. I.).

988.
[Anon.], Review of 'The Grenville Papers', *North British Review*, 19 (1853), 475–518.
(W. I.).

989–990.
[Anon.], Review of Weld's 'History of the Royal Society', *North British Review*, 20 (1853), 209–247; *Eclectic Magazine*, 31 (1854), 217–236.
(W. I., Poole).

991.
'On the Date of the Discovery of the Optical Properties of Chrysammate of Potash', [In a letter from Sir David Brewster to Professor Stokes] *London, Edinburgh and Dublin Philosophical Magazine and Journal of Science*, 4th series, 7 (1854), 171–172.
(Signed article).

992.
'Notice on Barometrical, Thermometrical, and Hygrometrical Clocks', *London, Edinburgh and Dublin Philosophical Magazine and Journal of Science*, 4th series, 7 (1854), 358.
(R. S. C. 264).

993–994.
[Anon.], 'François Arago—his Life and Discoveries', *North British Review*, 20 (1854), 459–500; *Eclectic Magazine*, 32 (1854), 145–166.
(W. I., Poole).

995.
[Anon.], 'Of the Plurality of Worlds', *North British Review*, 21 (1854), 1–44.
(W. I.).

996.
[Anon.], Review of Sir Roderick Murchison's 'Siluria', *North British Review*, 21 (1854), 505–544.
(W. I.).

997.
[Anon.], 'Sir H. Holland on Mental Physiology, Electro-Biology, etc.', *North British Review*, 22 (1854), 179–224.
(W. I.).

998.
[Anon.], 'The British Association for the Advancement of Science', *Hogg's Instructor*, new series, 3 (1854), 408–427. (Quotes Brewster extensively).

999–1009.
More Worlds than One; the creed of the philosophical and the hope of the Christian. (London, 1854).
—Third thousand (London, 1854).
—Fourth thousand, corrected and greatly enlarged (London, 1854).
—Another edition (New York, 1854).
—Another edition (New York, 1856).
—Another edition (New York, 1860).
—Ninth thousand (London, 1870).
—Another edition (New York, 1871?).
—New edition with illustrations (London, 1874).
—Another edition (London, 1876).
—Another edition (London, 1895).
(B. M. Cat., Nat. Un. Cat.).

1010–1011.
[Anon.], 'The Electric Telegraph', *North British Review*, 22 (1855), 545–591; *Eclectic Magazine*, 34 (1855), 466–484.
(W. I., Poole).

1012–1013.
[Anon.], Review of Muirhead's 'Life and Inventions of James Watt', *North British Review*, 23 (1855), 193–231; *Littell's Living Age*, 45 (1855), 743–762.
(W. I., Poole).

1014.
[Anon.], Review of Dr. Peacock's 'Life of Dr Thomas Young', *North British Review*, 23 (1855), 481–520.
(W. I.).

1015.
[Anon.], 'The Paris Exposition and the Patent Laws', *North British Review,* 24 (1855), 231–267.
(W. I.).

1016.
[Anon.], 'Scientific Pickings from the British Association Papers', *Hogg's Instructor,* new series, 5 (1855), 318–336. (Brewster's optical work reported in detail).

1017–1022.
Memoirs of the Life, Writings and Discoveries of Sir Isaac Newton, 2 volumes. (Edinburgh, 1855).
—second edition (Edinburgh, 1860).
—a new edition revised by W. T. Lyon (London, 1875).
A Short Life of Sir Isaac Newton (compiled from Sir David Brewster's work on the subject) [the editor's introduction signed MMC] (London, 1864).
—*Newton*... Im Auszuge und mit Anmerkungen zum Schulgebrauche herausgegeben von Dr E Schenk und Dr L Bahlsen... Mit einern Bildnis Newtona, etc. (Berlin, 1895).
—Reprinted from 1855 edition, Introduction by Richard S. Westfall (New York and London, 1965).
(B. M. Cat., Books in Print).

1023.
'On the Triple Spectrum', *Report of the British Association for the Advancement of Science... 1855* (London, 1856), Part II, 7–9.
(R. S. C. 266).

1024.
'On the Binocular Vision of Surfaces of Different Colours', *Report of the British Association for the Advancement of Science... 1855* (London, 1856), Part II, 9.
(R. S. C. 265).

1025.
'On the Absorption of Matter by the Surfaces of Bodies', *Report of the British Association for the Advancement of Science... 1855* (London, 1856), Part II, 9.
(R. S. C. 268).

1026.
'On the Binocular Vision of Surfaces of Different Colours', *Report of the British Association for the Advancement of Science... 1855* (London, 1856), Part II, 9.
(R. S. C. 265).

1027.
'On the Existence of Acari in Mica', *Report of the British Association for the Advancement of Science... 1855* (London, 1856), Part II, 9.
(R. S. C. 267).

1028.
'On the Remains of Plants in Calcareous Spar from King's Country, Ireland', *Report of the British Association for the Advancement of Science... 1855* (London, 1856), Part II, 9. (Signed article).

1029.
'On the Phenomena of Decomposed Glass', *Report of the British Association for the Advancement of Science... 1855* (London, 1856), Part II, 10. (Signed article).

1030.
[Anon.], 'Dr George Wilson on Colour-Blindness', *North British Review,* 24 (1856), 325–358.
(W. I.).

1031–1032.
[Anon.], 'The Weather and its Prognostics', *North British Review,* 25 (1856), 173–204; *Eclectic Magazine,* 38 (1856), 522–534.
(W. I., Poole).

1033.
[Anon.], 'The Microscope and its Revelations', *North British Review,* 25 (1856), 437–476.
(W. I.).

1034.
[Anon.], 'The Sight and How to See', *North British Review,* 26 (1856), 145–184.
(W. I.).

1035–1039.
The Stereoscope: its History, Theory and Construction, with its Application to the Fine and Useful Arts, and to Education etc., (London, 1856).
—Another edition (London, 1870).
—*Das Stereoskop; seine Geschichte, Theorie und Construction, nebst seiner Anwendung auf die schönen und nützlichen Künste und für die Zwecke des Jugenduntervichtes.* Aus dem Englischen von Dr. Christ. Heinr. Schmidt (Weimar, 1857).
—Another edition. Nebst einem Auhang über die Photographie der Stereoskopbilder nach de La Blanchère. Ins Deutsche übertrapen von Dr. Christ. Heinr. Schmidt... (Weimar, 1862).
—Reprint, Introduction by Rudolf Kingslake (London, 1971).
(B. M. Cat., Nat. Un. Cat., Books in Print).

1040–1041.
[Anon.], Review of Dr. Kane's 'Arctic Explorations', *North British Review,* 26 (1857), 407–442; *Eclectic Magazine,* 40 (1857), 433–454.
(W. I., Poole).

1042.
[Anon.], Review of 'Memoirs of John Dalton', *North British Review,* 27 (1857), 465–497.
(W. I.).

1043.
Paris Universal Exhibition: Report on certain Optical and other Instruments. (London, 1857). (Signed article, family volume).

1044.
'On the Centering of the Lenses of the Compound Object Glasses of Microscopes', *Report of the British Association for the Advancement of Science... 1857* (London, 1858), Part II, 4–5.
(R. S. C. 269).

1045.
'On the Optics of Photography, but Particularly on the Character of the Images formed upon Opaque and Transparent Surfaces', *Journal of the Photographic Society of London,* 4 (1858), 83–89.
(R. S. C. 275).

1046.
'Account of a new Photographic Process by M. Dupius', *Journal of the Photographic Society of London,* 4 (1858), 88.
(R. S. C. 276).

1047–1048.
[Anon.], Review of 'Rambles of a Naturalist etc.', *North British Review,* 28 (1858), 158–190; *Eclectic Magazine,* 44 (1858), 39–58.
(W. I., Poole).

1049–1050.
[Anon.], Review of Lieutenant Maury's 'Geography of the Sea', *North British Review,* 28 (1858), 403–436; *Eclectic Magazine,* 44 (1858), 433–453.
(W. I., Poole).

1051–1053.
[Anon.], 'Researches on Light—Sanatory—Scientific and Aesthetical', *North British Review,* 29 (1858), 178–210; *Littell's Living Age,* 59 (1858), 243–263; *Eclectic Magazine,* 45 (1858), 291–311.
(W. I., Poole).

1054.
[Anon.], 'The Atlantic Telegraph', *North British Review,* 29 (1858), 519–555.
(W. I.).

1055.
'On the Duration of Luminous Impressions on certain points of the Retina', *Report of the British Association for the Advancement of Science . . . 1858* (London, 1859), Part II, 6–7.
(R. S. C. 270).

1056.
'On Vision through the Foramen Centrale of the Retina', *Report of the British Association for the Advancement of Science . . . 1858* (London, 1859), Part II, 7.
(R. S. C. 271).

1057.
'On certain Abnormal Structures in the Crystalline Lenses of Animals, and in the Human Crystalline', *Report of the British Association for the Advancement of Science . . . 1858* (London, 1859), Part II, 7–10.
(R. S. C. 272).

1058.
'On the Crystalline Lens of the Cuttle-fish', *Report of the British Association for the Advancement of Science . . . 1858* (London, 1859), Part II, 10–13.
(R. S. C. 273).

1059.
'On the Use of Amethyst Plates in experiments on the Polarisation of Light', *Report of the British Association for the Advancement of Science . . . 1858* (London, 1859), Part II, 13.
(R. S. C. 274).

1060–1061.
'Note sur la polarisation de la lumière des comètes', *Comptes Rendus Hebdomadaires des Séances de l' Académie des Sciences, 48* (1859), 384–385; *London, Edinburgh and Dublin Philosophical Magazine and Journal of Science*, 4th series, *17* (1859), 359–360.
(R. S. C. 280).

1062–1064.
'On the Coloured Houppes or Sectors of Haidinger', *London, Edinburgh and Dublin Philosophical Magazine and Journal of Science*, 4th series, *17* (1859), 323–326; *Comptes Rendus Hebdomadaires des Séances de l' Académie des Sciences, 48* (1859), 614–618; *Annalen der Physik und Chemie; von J. C. Poggendorff, 107* (1859), 346–352.
(R. S. C. 281).

1065.
'On the Lines of the Solar Spectrum', *Proceedings of the Royal Society of London, 10* (1859–60), 339–341.
(R. S. C. 282).

1066–1068.
[Anon.], Review of De la Rive's 'Electricity in Theory and Practice', *North British Review, 30* (1859), 160–201; *Eclectic Magazine, 47* (1859), 39–51, 200–214.
(W. I., Poole).

1069–1070.
[Anon.], Review of 'Select Memoirs of Port-Royal', *North British Review, 30* (1859), 492–531; *Littell's Living Age, 62* (1859), 202–225.
(W. I., Poole).

1071.
[Anon.], 'Glaciers', *North British Review, 31* (1859), 89–124.
(W. I.).

1072.
[Anon.], 'Life-boats—Lightning Conductors—Lighthouses', *North British Review, 31* (1859), 492–529.
(W. I.).

1073.
Introductory Address . . . on the Opening of Session 1859–60 [of the University of Edinburgh] (Edinburgh, 1859).
(B. M. Cat.).

1074.
Memorial on the new system of dioptic lights invented and introduced by Sir David Brewster . . . to the Right Honourable the Lords Commissioners of Her Majesty's Treasury. (Cupar, 1859).
(B. M. Cat.).

1075.
[Anon.], 'Social Claims and Aspects of Science', *Meliora, 1* (1859), 237–259. (Brewster's rhetoric and style).

1076.
'On a New Species of Double Refraction', *Report of the British Association for the Advancement of Science . . . 1859* (London, 1860), Part II, 10–11.
(R. S. C. 277).

1077–1078.
'On the Decomposed Glass found at Nineveh and Other Places', *Report of the British Association for the Advancement of Science . . . 1859* (London, 1860), Part II, 11; *ibid.*, . . . *1860* (London, 1861), Part II, 9–12.
(R. S. C. 278).

1079.
'On Sir Christopher Wren's Cipher, containing Three Methods of Finding the Longitude', *Report of the British Association for the Advancement of Science . . . 1859* (London, 1860), Part II, 34–35.
(R. S. C. 279).

1080.
'On a Horseshoe Nail found in the Red Sandstone of Kingoodie', *Report of the British Association for the Advancement of Science . . . 1859* (London, 1860), Part II, 101. (Signed article).

1081.
'On a Remarkable Specimen of Chalcedony, belonging to Miss Campbell, and Exhibiting a perfectly Distinct and Well-drawn Landscape', *Report of the British Association for the Advancement of Science . . . 1859* (London, 1860), Part II, 245. (Signed article).

1082.
'On the Connection between Solar Spots and Magnetic Disturbances', *Report of the British Association for the Advancement of Science . . . 1859* (London, 1860), Part II, 245. (Signed article).

1083.
'Observations sur un point d'histoire de l'optique', *Comptes Rendus Hebdomadaires des Séances de l' Académie des Sciences, 51* (1860), 425–429, 467.
(R. S. C. 285).

1084.
'On the Invention of the Stereoscope in the 16th century, and of Binocular Drawings, by Jacopo Chimenti da Empoli', *Journal of the Photographic Society of London, 6* (1860), 232–233.
(R. S. C. 286).

1085–1086.
Reply to Messrs. D. & T. Stevenson's Pamphlet on Lighthouses (Edinburgh, 1860).
[—see David & T. Stevenson *Answer to Sir David Brewster's Reply to Messrs. Stevenson's Pamphlet on Sir D. Brewster's Memorial to the Treasury*, 1860.]
—*A new edition . . . enlarged, with a preface and original documents* (Edinburgh, 1860).
(B. M. Cat.).

1087.
Introductory Address . . . at the Opening of Session, 1860–61 [of the University of Edinburgh] (Edinburgh, 1860).
(B. M. Cat.).

1088.
[Anon.], 'Form and Colour—Sir G. Wilkinson', *North British Review, 32* (1860), 126–158.
(W. I.).

1089.
[Anon.], 'British Lighthouses', *North British Review*, 32 (1860), 487–519.
(W. I.).

1090–1091.
[Anon.], 'Romance of the New Planet', *North British Review*, 33 (1860), 1–20; *Eclectic Magazine*, 51 (1860), 351–358.
(W. I., Poole).

1092.
[Anon.], 'Recent Theories in Meteorology', *North British Review*, 33 (1860), 256–267.
(W. I.).

1093–1095.
[Anon.], 'The Martyrdom of Galileo', *North British Review*, 33 (1860), 513–548; *Eclectic Magazine*, 52 (1861), 199–209, 303–314.
(W. I., Poole).

1096.
D. B., 'Electricity', *Encyclopaedia Britannica*, 21 vols., 8th edition, (Edinburgh, 1853–1860), VIII, 523–627.
(List).

1097.
D. B., 'Hydrodynamics' *Encyclopaedia Britannica*, 21 vols., 8th edition, (Edinburgh, 1853–1860), XII, 69–184.
(List).

1098.
[Anon.], 'Kaleidoscope', *Encyclopaedia Britannica*, 21 vols., 8th edition, (Edinburgh, 1853–1860), XIII, 37–40.
(List).

1099.
D. B., 'Magnetism' *Encyclopaedia Britannica*, 21 vols., 8th edition, (Edinburgh, 1853–1860), XIV, 1–92.
(List).

1100.
D. B., 'Micrometer' *Encyclopaedia Britannica*, 21 vols., 8th edition, (Edinburgh, 1853–1860), XIV, 742–762.
(List).

1101.
D. B., 'Microscope' *Encyclopaedia Britannica*, 21 vols., 8th edition, (Edinburgh, 1853–1860), XIV, 763–806.
(List).

1102.
D. B., 'Newton' *Encyclopaedia Britannica*, 21 vols., 8th edition, (Edinburgh, 1853–1860), XVI, 205–210.
(List).

1103.
D. B., 'Optics' *Encyclopaedia Britannica*, 21 vols., 8th edition, (Edinburgh, 1853–1860), XVI, 512–618.
(List).

1104.
D. B., 'Photogalvanography' *Encyclopaedia Britannica*, 21 vols., 8th edition, (Edinburgh, 1853–1860), XVII, 544.
(List).

1105.
D. B., 'Photoglyphic engraving' *Encyclopaedia Britannica*, 21 vols., 8th edition, (Edinburgh, 1853–1860), XVII, 544.
(List).

1106.
D. B., 'Photography' *Encyclopaedia Britannica*, 21 vols., 8th edition, (Edinburgh, 1853–1860), XVII, 544–554.
(List).

1107.
D. B., 'Photo-lithography' *Encyclopaedia Britannica*, 21 vols., 8th edition, (Edinburgh, 1853–1860), XVII, 554–555.
(List).

1108.
D. B., 'Stereoscope' *Encyclopaedia Britannica*, 21 vols., 8th edition, (Edinburgh, 1853–1860), XX, 684–691.
(List).

1109.
D. B., 'Voltaic electricity' *Encyclopaedia Britannica*, 21 vols., 8th edition (Edinburgh, 1853–1860), XXI, 609–655.
(List).

1110.
'On the Influence of Very Small Apertures on Telescopic Vision', *Report of the British Association for the Advancement of Science . . . 1860* (London, 1861), Part II, 7. (Signed article).

1111.
'On some Optical Illusions connected with the Inversion of Perspective', *Report of the British Association for the Advancement of Science . . . 1860* (London, 1861), Part II, 7–8.
(R. S. C. 283).

1112–1113.
'On Microscopic Vision and a New Form of Microscope', *Report of the British Association for the Advancement of Science . . . 1860* (London, 1861), Part II, 8–9; *Journal of the Franklin Institute*, 70 (1860), 249.
(R. S. C. 284, Poole).

1114.
'On the Decomposed Glass Found at Nineveh and other Places', *Report of the British Association for the Advancement of Science . . . 1860*. (London, 1861), Part II, 9–12. (Signed article).

1115.
'Details Respecting a Nail Found in Kingoodie Quarry, 1843', *Report of the British Association for the Advancement of Science . . . 1860* (London, 1861), Part II, 73. (Signed article).

1116–1117.
'On the Action of Uncrystallised Films upon Common and Polarised Light', *Transactions of the Royal Society of Edinburgh*, 22 (1861), 607–610; *London, Edinburgh and Dublin Philosophical Magazine and Journal of Science*, 4th series, 22 (1861), 269–273.
(R. S. C. 288).

1118.
'On Certain Affections of the Retina', *London, Edinburgh and Dublin Philosophical Magazine and Journal of Science*, 4th series, 21 (1861), 20–24.
(R. S. C. 295).

1119.
[Anon.], 'Engineering and Engineers', *North British Review*, 34 (1861), 142–183.
(W. I.).

1120.
[Anon.], 'Railway Accidents', *North British Review*, 34 (1861), 399–427.
(W. I.).

1121.
[Anon.], Review of Du Chaillu's 'Explorations and Adventures', *North British Review*, 35 (1861), 219–252.
(W. I.).

1122.
[Anon.], 'Comets', *North British Review*, 35 (1861), 495–533.
(W. I.).

1123.
[Anon.], 'Alexander von Humboldt', *Meliora, 3* (1861), 273–295.
(*Cf.* 854, 866, 909 and 919).

1124.
'On Photographic Micrometers', *Report of the British Association for the Advancement of Science ... 1861* (London, 1862), Part II, 28. (Signed article).

1125.
'On the Compensation of Impressions moving over the Retina', *Report of the British Association for the Advancement of Science ... 1861* (London, 1862), Part II, 29. (Signed article).

1126.
'On the Optical Study of the Retina', *Report of the British Association for the Advancement of Science ... 1861* (London, 1862), Part II, 29. (Signed article).

1127.
'On Binocular Lustre', *Report of the British Association for the Advancement of Science ... 1861* (London, 1862), Part II, 29–31.
(R. S. C. 287).

1128.
'On the Action of Various Coloured Bodies on the Spectrum', *London, Edinburgh and Dublin Philosophical Magazine and Journal of Science*, 4th series, 24 (1862), 441–447.
(R. S. C. 297).

1129.
'On Photographic and Stereoscopic Portraiture', *Journal of the Photographic Society of London*, 7 (1862), 130–132, 190.
(R. S. C. 298).

1130.
'On Discoveries and Inventions in Photography', *Journal of the Photographic Society of London*, 7 (1862), 183.
(R. S. C. 299).

1131–1133.
'On the Stereoscopic Picture Executed in the 16th century by Jacopo Chimenti', *Journal of the Photographic Society of London*, 7 (1862), 9–11; *British Journal of Photography*, 9 (1862), 105–106; *London, Edinburgh and Dublin Philosophical Magazine and Journal of Science*, 4th series, 27 (1864), 33–35.
(R. S. C. 305).

1134.
'On Zincography', *British Journal of Photography*, 9 (1862), 72. [Extract from 1141]. (Signed article).

1135–1137.
The History of the Invention of the Dioptric Lights and their Introduction into Great Britain. (London, 1862).
—Another edition (1864).
—Another edition (1865).
(B. M. Cat., family volume).

1138.
'Facts and Fancies of Mr Darwin', *Good Words*, 3 (1862), 3–9. (Signed article).

1139.
'The Eye: its Structure and Powers', *Good Words*, 3 (1862), 170–176. (Signed article).

1140.
'The Human Eye: its Phenomena and Illusions', *Good Words*, 3 (1862), 498–503. (Signed article).

1141.
[Anon.], 'Recent Progress of Photographic Art', *North British Review*, 36 (1862), 170–203.
(W. I.).

1142–1143.
[Anon.], 'Sir G. C. Lewis on the Astronomy of the Ancients', *North British Review*, 36 (1862), 485–513; *Eclectic Magazine*, 56 (1862), 433–450.
(W. I., Poole).

1144.
[Anon.], Review of Professor Faivre's 'Scientific Biography of Goethe', *North British Review*, 38 (1863), 107–133.
(W. I.).

1145–1148.
[Anon.], 'Pretensions of Spiritualism—Life of D. D. Home', *North British Review*, 39 (1863), 174–206; *Littell's Living Age*, 78 (1863), 531–550; *Eclectic Magazine*, 60 (1863), 173–186; 291–297.
(W. I., Poole).

1149–1150.
'On the Photomicroscope', *British Journal of Photography*, 11 (1864), 22–23; *Journal of the Photographic Society of London*, 8 (1864), 439–441.
(R. S. C. 304).

1151.
'On the Stereoscopic Relief in the Chimenti Pictures', *British Journal of Photography*, 11 (1864), 33. (Signed article).

1152.
'Life in a Drop of Water', *Good Words*, 5 (1864), 169–176. (Signed article).

1153.
Introductory Address ... at the opening of session 1864–5. [At the University of Edinburgh.] (Edinburgh, 1864).
(EUL and family volume).

1154.
'Presidential Address', *Proceedings of the Royal Society of Edinburgh*, 5 (1864–65), 321–336.
(EUL).

1155–1156.
'On the Pressure Cavities in Topaz, Beryl and Diamond, and their bearing on Geological Theories', *Transactions of the Royal Society of Edinburgh*, 23 (1864), 39–44; *London, Edinburgh and Dublin Philosophical Magazine and Journal of Science*, 4th series, 25 (1863), 174–180.
(R. S. C. 289).

1157.
'On the Existence of Acari between the Laminae of Mica in Optical Contact', *Transactions of the Royal Society of Edinburgh*, 23 (1864), 95–96.
(R. S. C. 290).

1158.
'On certain Vegetable and Mineral Formations in Calcareous Spar', *Transactions of the Royal Society of Edinburgh*, 23 (1864), 97–98.
(R. S. C. 291).

1159.
'On the Structure and Optical Phenomena of Ancient Decomposed Glass', *Transactions of the Royal Society of Edinburgh*, 23 (1864), 193–204.
(R. S. C. 292).

1160–1161.
'On the Polarisation of Light by Rough and White Surfaces', *Transactions of the Royal Society of Edinburgh*, 23 (1864), 205–210; *London, Edinburgh and Dublin Philosophical Magazine and Journal of Science*, 4th series, 25 (1863), 344–350.
(R. S. C. 293).

1162–1164.
'Observations on the Polarisation of the Atmosphere made at St. Andrews in 1841, 1842, 1843, 1844 and 1845', *Transactions of the*

Royal Society of Edinburgh, 23 (1864), 211–240; *London, Edinburgh and Dublin Philosophical Magazine and Journal of Science*, 4th series, 30 (1865), 118–129, 161–181.
(R. S. C. 294).

1165.
'Description of the Lithoscope, an Instrument for Distinguishing Precious Stones and Other Bodies', *Transactions of the Royal Society of Edinburgh*, 23 (1864), 419–425.
(R. S. C. 296).

1166.
'Sur les observations et les correspondances météorologiques établies en Écosse', *Comptes Rendus Hebdomadaires des Séances de l'Académie des Sciences*, 61 (1865), 97–100.
(R. S. C. 10).

1167–1168.
'On the Cause and Cure of Cataract', *London, Edinburgh and Dublin Philosophical Magazine and Journal of Science*, 4th series, 29 (1865), 426–430; *Transactions of the Royal Society of Edinburgh*, 24 (1867), 11–14.
(R. S. C. 11).

1169–1171.
'On Hemiopsy, or Half-vision', *London, Edinburgh and Dublin Philosophical Magazine and Journal of Science*, 4th series, 29 (1865), 503–507; *Proceedings of the Royal Society of Edinburgh*, 5 (1866), 373; *Transactions of the Royal Society of Edinburgh*, 24 (1867), 15–18.
(R. S. C. 12).

1172.
'Observations on the previous paper', [about Chinese Magic Mirrors], *The Scientific Review and Journal of the Inventors' Institute*, 1 (1865), 20–21. (Signed article).

1173.
'Additional observations', [as above], *The Scientific Review and Journal of the Inventors' Institute*, 1 (1865), 34. (Signed article).

1174.
'On the Polarisation of the Atmosphere', *The Scientific Review and Journal of the Inventors' Institute*, 1 (1865), 81–82. (Signed article).

1175.
'Presidential Address', *Proceedings of the Royal Society of Edinburgh*, 6 (1865–66), 458–483. (Signed article).

1176.
'On the Patent Laws', *The Scientific Review and Journal of the Inventors' Institute*, 1 (1866), 169–170. (Signed article).

1177–1180.
'On the Bands Formed by the Superposition of Paragenic Spectra produced by the Grooved Surfaces of Glass and Steel', *London, Edinburgh and Dublin Philosophical Magazine and Journal of Science*, 4th series, 31 (1866), 22–26, 98–104; *Transactions of the Royal Society of Edinburgh*, 24 (1867), 221–225, 227–232.
(R. S. C. 13).

1181–1182.
'On the Influence of the Doubly Refracting Force of Calcareous Spar on the Polarisation, the Intensity, and the Colour of the Light which it Reflects', *Proceedings of the Royal Society of Edinburgh*, 5 (1866), 175–176; *Transactions of the Royal Society of Edinburgh*, 24 (1867), 233–246.
(R. S. C. 14).

1183.
'On the Defraction Bands produced by Double Striated Surfaces', *Proceedings of the Royal Society of Edinburgh*, 5 (1866), 184.
(R. S. C. 15).

1184–1185.
'On a New Property of the Retina', *Proceedings of the Royal Society of Edinburgh*, 5 (1866), 373–374; *Transactions of the Royal Society of Edinburgh*, 24 (1867), 327–330.
(R. S. C. 16).

1186.
'On the Fairy Stones found in the Elwand Water near Melrose', *Proceedings of the Royal Society of Edinburgh*, 5 (1866), 567–571.
(R. S. C. 17).

1187.
'Notes sur l'histoire de l'analyse spectrale', *Comptes Rendus Hebdomadaires des Séances de l'Académie des Sciences*, 62 (1866), 17–19.
(R. S. C. 18).

1188.
'On the Rights of Inventors and Authors', *The Scientific Review and Journal of the Inventors' Institute*, 2 (1866), 65. (Signed article).

1189.
'On the Methods of taking Transparent Photographs of Opaque Objects', *The Scientific Review and Journal of the Inventors' Institute*, 2 (1866), 73–74. (Signed article).

1190–1191.
'On the claims of Science, Literature and the Arts of National Recognition and Support', *The Scientific Review and Journal of the Inventors' Institute*, 2 (1866), 101–102, 117–118. (Signed article).

1192–1193.
'On the Effects of Light', *The Scientific Review and Journal of the Inventors' Institute*, 2 (1867), 197–198, 216–217. (Signed article).

1194.
'On the Equilibrium of Liquid Films', *The Scientific Review and Journal of the Inventors' Institute*, 2 (1867), 234. (Signed article).

1195.
'On the Vapour Lines in the Spectrum', *The Scientific Review and Journal of the Inventors' Institute*, 2 (1867), 284. (Signed article).

1196–1198.
'On the Colours of the Soap-bubble' *Transactions of the Royal Society of Edinburgh*, 24 (1867), 491–504; *Report of the British Association for the Advancement of Science ... 1867* (London, 1868) Part II, 6; *Proceedings of the Royal Society of Edinburgh*, 6 (1869), 64–65.
(R. S. C. 19).

1199–1202.
'On the Figures of Equilibrium of Liquid Films' *Transactions of the Royal Society of Edinburgh*, 24 (1867), 505–513; *Report of the British Association for the Advancement of Science ... 1867* (London, 1868) Part II, 6–8; *Proceedings of the Royal Society of Edinburgh*, 6 (1869), 76–78, 311–313.
(R. S. C. 20).

1203–1205.
'On the Motions and Colours upon Films of Alcohol, Volatile Oils, and Other Fluids', *Transactions of the Royal Society of Edinburgh*, 24 (1867), 653–656; *Report of the British Association for the Advancement of Science ... 1867* (London, 1868) Part II, 8; *Proceedings of the Royal Society of Edinburgh*, 6 (1869), 108–109.
(R. S. C. 21).

1206–1210.
'On the Radiant Spectrum', *London, Edinburgh and Dublin Philosophical Magazine and Journal of Science*, 4th series, 34 (1867), 202–204; *Report of the British Association for the Advancement of Science ... 1867* (London, 1868) Part II, 8–10; *British Journal of Photography*, 14 (1867), 430–439; *Annales de Chemie ou Recueil de Mémoires concernant la Chimie et les Arts qui en dépendent*, 13 (1868), 467–468; *Proceedings of the Royal Society of Edinburgh*, 6 (1869), 147–149.
(R. S. C. 22).

1211–1214.
'Additional Observations on the Polarisation of the Atmosphere made at St. Andrews in 1841, 1842, 1843, 1844, and 1845',

Transactions of the Royal Society of Edinburgh, 24 (1867), 247–286; *London, Edinburgh and Dublin Philosophical Magazine and Journal of Science,* 4th series, *33* (1867) 290–304, 346–360, 455–456.
(R. S. C. 23).

1215.
'Report on the Hourly Meteorological Register kept at Leith Fort in the years 1826 and 1827', *Transactions of the Royal Society of Edinburgh, 24* (1867), 351–362.
(R. S. C. 24).

1216.
'Description of a Double Holophote Apparatus for Lighthouses, and of a Method of Introducing the Electric or other Lights', *Transactions of the Royal Society of Edinburgh, 24* (1867), 635–648.
(R. S. C. 25).

1217.
'Sur la prétendue correspondance entre Newton et Pascal', *Comptes Rendus Hebdomadaires des Séances de l'Académie des Sciences, 65* (1867), 261–263.
(R. S. C. 26).

1218.
'Lettre à M. Chevreul au sujet des lettres attribuées à Pascal et à Newton,' *Comptes Rendus des Hebdomadaires des Séances de l'Académie des Sciences, 65* (1867), 537–538.
(R. S. C. 27).

1219.
'Nouvelle lettre à M. Chevreul, au sujet des rapports qui auraient existé entre Newton et Pascal', *Comptes Rendus Hebdomadaires des Séances de l'Académie des Sciences, 65* (1867), 653–655.
(R. S. C. 28).

1220.
'Lettre à M. Chevreul, au sujet des relations qui auraient existé entre Pascal et Newton', *Comptes Rendus Hebdomadaires des Séances de l'Académie des Sciences, 65* (1867), 717–718.
(R. S. C. 29).

1221.
'Lettre à M. Le Verrier, concernant les relations qui ont existé entre Jacques Cassini et Newton', *Comptes Rendus Hebdomadaires des Séances de l'Académie des Sciences, 65* (1867), 769–770.
(R. S. C. 30).

1222.
'Lettre concernant les documents attribués à Pascal et à Newton', *Comptes Rendus Hebdomadaires des Séances de l'Académie des Sciences, 65* (1867), 770–771.
(R. S. C. 31).

1223.
'Lettre à M. Chevreul, au sujet de l'authenticité des pièces attribuées à Pascal et à Newton', *Comptes Rendus Hebdomadaires des Séances de l'Académie des Sciences, 65* (1867), 825–826.
(R. S. C. 32).

1224.
'Lettre adressée à M. Chevreul, au sujet des pièces relatives à Newton et à Pascal, pièces considérées comme provenant de la collection de Desmaizeaux', *Comptes Rendus Hebdomadaires des Séances de l'Académie des Sciences, 65* (1867), 925–926.
(R. S. C. 33).

1225–1226.
'Notice respecting the Enamel Photographs executed by Mr. McRaw of Edinburgh', *British Journal of Photography, 14* (1867), 439; *Report of the British Association for the Advancement of Science ... 1867* (London, 1868) Part II, 8. (Signed article).

1227–1228.
'Scientific Education in our Schools', *The Scientific Review and Journal of the Inventors' Institute, 3* (1868), 43–44, 62–63. (Signed article).

1229.
'On the Alleged Correspondence between Pascal and Newton', *Report of the British Association for the Advancement of Science ... 1867* (London, 1868) Part II, 1–2. (Signed article).

1230.
'Notice respecting a Haystack struck by Lightning', *Report of the British Association for the Advancement of Science ... 1867* (London, 1868) Part, 19. (Signed article).

1231.
'Presidential Address', *Proceedings of the Royal Society of Edinburgh, 6* (1869), 2–36.
(R. S. C. 34).

1232.
'On the Vapour Lines of the Spectrum', *Proceedings of the Royal Society of Edinburgh, 6* (1869), 145–147.
(R. S. C. 35).

1233.
'Presidential Address', *Proceedings of the Royal Society of Edinburgh, 6* (1869), 174–185.
(R. S. C. 36).

1234.
'On the Motion, Equilibrium and Forms of Liquid Films', *Transactions of the Royal Society of Edinburgh, 25* (1869), 111–118.
(R. S. C. 37).

Joint Papers

1235. David Brewster and J. H. Gladstone 'On the lines of the solar spectrum'. *Philosophical Transactions of the Royal Society of London, 150* (1860), 149–160.

1236–1237. David Brewster and John Gordon 'Experiments on the structure and refractive power of the coats and humours of the human eye'. *Edinburgh Philosophical Journal, 1* (1819), 42–45; *Annales de Chimie, ou Receul de Mémoires concernant la Chemie et les Arts qui en dependent, 11* (1812), 330–332.

1238–1239. David Brewster and Leonard Horner 'On an artificial substance resembling shell; with an account of an examination of the same', *Philosophical Transactions of the Royal Society of London, 126* (1836), 49–56; *London and Edinburgh Philosophical Magazine and Journal of Science,* 3rd series, *10* (1837), 201–210.

1240. David Brewster and T. J. Seebeck 'Sur le développement des forces polarisantes par la pression'. *Nouveau Bulletin des Sciences de la Société Philomathique de Paris,* (1816), 49–51.

1241. David Brewster and J. C. von Yelin 'Eine neue in den Höhlungen von Mineralien entdeckte Flüssigkeit von sonderbaren physikalischen Eigenschaften'. *Annalen der Physik; von L. W. Gilbert 74* (1823), 331–333.

Appendix II(c)

Published Writings about Sir David Brewster

A. D. MORRISON-LOW

1. George B. Airy
'Remarks on Sir David Brewster's paper "On the Absorption of Specific Rays etc."', *London and Edinburgh Philosophical Magazine and Journal of Science*, 3rd series, 2 (1833), 419–424.

2. [Anon.]
'David Brewster', *Biographie Universelle et Portative des Contemporaines*, (Paris, 1836), V, 77–81.

3. John Fearn
An Appeal to Philosophers by name, on demonstration of vision in the brain, and against the attack by Sir David Brewster on the rationale of cerebral vision [With special reference to Brewster's *Letters on Natural Magic*]. (London, 1837).

4. [? Hugh Miller]
'Libel against Sir David Brewster', *The Witness, 5* (6 January 1844).

5. Rev. Anstruther Taylor and John Inglis
'Correspondence in Sir David Brewster's case', *The Witness, 5* (6 April 1844).

6. Presbytery of St. Andrews
Case for the Presbytery of St. Andrews, in the libel against Sir D. Brewster (Edinburgh, 1845).

7. Presbytery of St. Andrews
The Grievance of University Tests as set forth in the Proceedings of the Presbytery of St. Andrews. With an authentic copy of the libel in the case of Sir D. Brewster (Edinburgh, 1845).

8. George B. Airy
'On Sir David Brewster's new analysis of solar light', *London, Edinburgh and Dublin Philosophical Magazine and Journal of Science*, 3rd series, 30 (1847), 73–76.

9. Hermann von Helmholtz
'On Sir David Brewster's new analysis of solar light', *London, Edinburgh and Dublin Philosophical Magazine and Journal of Science*, 4th series, 4 (1852), 401–416.

10. [Augustus De Morgan]
Review of Sir David Brewster's 'Life of Newton', *North British Review*, 23 (1855), 261–264, 415.

11. David and Thomas Stevenson
Answer to Sir David Brewster's Reply to Messrs. Stevenson's Pamphlet on Sir D. Brewster's Memorial to the Treasury (Edinburgh, 1860).

12. J. C. Poggendorff
'Sir David Brewster', *Biographisch—Literarisches Handwörterbuch zur Geschichte der Exacten Wissenschaften* 2 vols., (Leipzig, 1863), I, col. 295–297.

13. C. Haushofer
Ueber den Asterismus und die Brewsters'chen Lichtfiguren am Calcite (Munich, 1865).

14. Lyon Playfair, James Y. Simpson and D. Stevenson
'Sir David Brewster', *Proceedings of the Royal Society of Edinburgh*, 6 (1866–1869), 282–284.
—*Sir David Brewster: His Last Days and Death ... at the Meeting of the Royal Society of Edinburgh 17 February 1868* (Edinburgh, 1868).

15. [Anon.]
'Sir D. Brewster', *Gentlemen's Magazine*, 4th series, 5 (1868), 539–541.

16. [Anon.]
'Sir David Brewster', *American Journal of Pharmacy*, 40 (1868), 287.

17. [Anon.]
'Sir David Brewster', *American Journal of Science*, 95 (1868), 284.

18. John C. Geikie
Michael Faraday and Sir David Brewster; Philosophers and Christians: Lessons from their Lives (London, 1868).

19. J. H. G[ladstone]
'Sir David Brewster', *Proceedings of the Royal Society of London*, 17 (1869), lxix–lxxiv.

20. [Anon.]
'Sir David Brewster', *Popular Science Monthly*, 26 (1854–85), 546–552.

21. Margaret Maria Gordon
The Home Life of Sir David Brewster (Edinburgh, 1869).
—2nd edition 1870.
—3rd edition 1881.

22. J. H. Gladstone
Review of 'The Home Life', *Nature, 1* (1870), 650.

23. [Anon.]
'Sir David Brewster', *Proceedings of the American Academy of Arts and Science, 8* (1873), 38–43.

24. J[ohn] D[uns]
'Sir David Brewster', in [Anon.] *Disruption Worthies* (Edinburgh, 1876), 195–200.

25. [Anon.]
'Sir David Brewster's Scientific Work', *Nature, 25* (1881), 157–159.

26. R[obert] H[unt]
'Sir David Brewster', in Leslie Stephen (ed.), *Dictionary of National Biography*, 63 vols., (London, 1886), VI, 299–303.

27. John G. Burke
'The Concepts of Crystal Symmetry', in *Origins of The Science of Crystals* (Berkeley, 1965), 147–175.

28. Richard S. Westfall
'Introduction', to David Brewster *Memoirs of Sir Isaac Newton*, 2 vols., (1855; reprinted New York and London, 1965) I, ix–xlv.

29. John R. Levene
'Sir David Brewster, and the Measurement of Refractive "Power"', *Proceedings of the Royal Microscopical Society, 1* (1966), 71–74.

30. Arthur T. Gill
'Sir David Brewster (1781–1868)', *The Photographic Journal, 108* (1968), 40–42.

31. Edgar W. Morse
'David Brewster', in C. C. Gillispie (ed.), *Dictionary of Scientific Biography*, 16 vols., (New York, 1970–1976), II, 451–454.

32. Herbert Rex Harvey
'An Inquiry into the Contribution to Science and the Advancement of Science of Sir David Brewster, KH, DCL. (1781–1868)'. Unpublished MA dissertation, University of Exeter, 1972.

33. Edgar W. Morse
'Natural Philosophy, Hypotheses and Impiety: Sir David Brewster confronts the Undulatory Theory of Light'. Unpublished Ph.D. thesis, University of California, Berkeley, 1972.

34. M. A. Sutton
'Sir John Herschel and the Development of Spectroscopy in Britain', *British Journal for the History of Science, 7* (1974), 42–60.

35. John Hedley Brooke
'Natural Theology and the Plurality of Worlds: Observations on the Brewster-Whewell Debate', *Annals of Science, 34* (1977), 221–286.

36. Henry John Steffens
'The Tenacity of Newtonian Optics in England: David Brewster, the Last Champion', in *The Development of Newtonian Optics in England* (New York, 1977), 137–151.

37. S. S. Schweber
'The Origin of the *Origin* Revisited', *Journal of the History of Biology, 10* (1977), 229–316.

38. William Cochran
'Who Remembers David Brewster?' *New Scientist, 92* (1981), 815–817.

39. John Forsyth
'The Mind of David Brewster', *The Scots Magazine*, new series *116* (1981–82), 287–295.

40. Paul D. Sherman
'Spectral Analysis by Coloured Filters and Brewster's Theory of the Spectrum' and 'Hermann von Helmholtz and the Demise of Brewster's Theory', in *Colour Vision in the Nineteenth Century: the Young-Helmholtz-Maxwell Theory* (Bristol, 1981), 20–59.

41. G. N. Cantor
Optics after Newton: Theories of Light in Britain 1704–1740 (Manchester, 1983).

42. J. Lindgren-Harley (ed.)
The Correspondence of Sir David Brewster to William Henry Fox Talbot Esq., 1831–1865 (? London, *forthcoming*).

43. A. D. Morrison-Low
'Sir David Brewster 1781–1868', in 'Four Centuries of Graduates', *University of Edinburgh Journal, 31* (1983), 49–51.

44. A. D. Morrison-Low
'The Origins of the Polarising Microscope: Sir David Brewster *versus* William Nicol', *(forthcoming)*.

45. Paul D. Sherman
Sir David Brewster: Natural Philosopher (Bristol, *forthcoming*).

46. A. D. C. Simpson
'Brewster's Society of Arts and the Pantograph Dispute'. *(forthcoming)*.

47. Nicholas Wade (ed.)
Brewster and Wheatstone on Vision (London, 1983).

48. Frank A. J. L. James
'The Debate on the Nature of Absorption of Light, 1830–1835: A Core-set Analysis' *History of Science, (forthcoming)*.